Lecture Notes
in Control and Information Sciences   368

Editors: M. Thoma, M. Morari

Frederick Chee, Tyrone Fernando

Closed-Loop Control of Blood Glucose

 Springer

Authors

Frederick Chee

School of Electrical
Electronic and Computer Engineering
University of Western Australia
35 Stirling Highway
Crawley, WA 6009
Australia
Email: hf.chee@ieee.org

Tyrone Fernando

School of Electrical
Electronic and Computer Engineering
University of Western Australia
35 Stirling Highway
Crawley, WA 6009
Australia
Email: tyrone@ee.uwa.edu.au

Library of Congress Control Number: 2007932674

ISSN print edition: 0170-8643
ISSN electronic edition: 1610-7411
ISBN-10 3-540-74030-9 Springer Berlin Heidelberg New York
ISBN-13 978-3-540-74030-8 Springer Berlin Heidelberg New York

Springer is a part of Springer Science+Business Media
springer.com
© Springer-Verlag Berlin Heidelberg 2007

Typesetting: by the authors and SPS using a Springer LATEX macro package

Printed on acid-free paper   SPIN: 12077946   89/SPS   5 4 3 2 1 0

To Jehovah God, my parents and siblings
Frederick

To my wife Yasmin and two daughters Melissa and Rochelle
Tyrone

Preface

Diabetes is a disease that is now regarded an epidemic in the world and a significant effort is directed towards finding better ways to manage diabetes. Keeping blood glucose levels as close to normal as possible, leads to a substantial decrease in long term complications of diabetes and can bring significant cost reductions associated with the disease. Traditionally, managing diabetes has been through intermittent monitoring of blood glucose and then administering an appropriate dose of insulin into the blood stream. This method of intermittent monitoring and administration of insulin cannot ensure blood glucose remains at near normal levels at all times and therefore, there is considerable interest in managing diabetes on a continuous basis.

The development of artificial organs/apparatus that regulate human's blood glucose level has been in progress since 1960. The aim was to measure blood glucose level ex vivo and then injecting an appropriate amount of insulin to the hyperglycaemic patient, thereby correcting the high glucose level. This aim of closing the "loop" is still being challenged by technological barriers even today, and progress are being made constantly both in overcoming the challenges and understanding more about the workings of glucose-regulatory system.

The purpose of this book is to introduce the field of closed-loop blood glucose control, in a simple manner, to the reader. This includes the hardware and software components that make up the control system (see Chapter 2). The hardware components involved the different types of glucose sensor (invasive, minimally-invasive and non-invasive) and the different types of insulin. The software component represents the approaches that translate a given blood glucose reading to an appropriate insulin rate (see Chapter 4). Some of these approaches applied mathematical sciences to model the underlying system and formulate mathematical solutions. Examples of how mathematical models are formulated as well as the control algorithms that stem from mathematical exercises are given, where possible.

This book also attempts to describe, from functional level, the basic physiology of blood glucose regulation during fasting and meal (see Chapter 3). The physiological changes in diabetic and critically ill patients are also described.

This would help in understanding the nature of the glucose-regulatory problem that we are facing and tackling. Finally, an example design of a closed-loop hardware using commercially-available equipment is given (see Chapter 5).

The ultimate goal in closed-loop control of blood glucose is not just finding the optimal insulin rates that can effectively reduce the high blood glucose, but to infuse it in such a way that the blood glucose level can mimic the body's natural excursion.

Perth   *Frederick Chee*
June 2007   *Tyrone Fernando*

Contents

1

Introduction

1.1 Introduction

Hyperglycaemia refers to an elevated glucose concentration in the circulating blood. While blood glucose level (BG) is often elevated after a meal, it usually normalises to a range of 3.5–5.6 mmol/l within 3 hours of a meal in a healthy individual. During fasting, BGs are also usually maintained within the normal range of 3.5–5.6 mmol/l [1].

Patients with diabetes mellitus, however, exhibit impaired regulatory responses. BG remains elevated postprandially, and during the fasting state. Patients under surgical or medical stress may also exhibit a diabetogenic response, even though they are not diabetics [2]. This is due to an elevated catecholamine and other "counter-regulatory" hormones in their circulation.

1.2 Causes of Hyperglycaemia

1.2.1 Physiological Background

Blood glucose level is usually regulated by two hormones – insulin and glucagon – secreted by the endocrine pancreas. Insulin is anabolic, and causes rapid uptake and use of glucose by most tissues in the body. It also causes the storage of excess glucose as glycogen, mainly in the liver and skeletal muscles. Excess glucose, that cannot be stored as glycogen, is converted to fatty acids and stored in adipose tissues. Insulin also promotes protein synthesis and storage.

Conversely, glucagon is catabolic, mobilizing glucose, fatty acids and amino acids from stores to the circulation, primarily through a breakdown of liver glycogen (glycogenolysis) and the generation of glucose from amino acids (gluconeogenesis).

The two hormones (insulin and glucagon) are reciprocal in their overall action and are secreted appropriately in most circumstances to keep the blood glucose concentration within the normal range [3]. When the glucose concentration rises too high, insulin is secreted, which then lowers the blood glucose concentration

F. Chee & T. Fernando: Closed-Loop Control of Blood Glucose, LNCIS 368, pp. 1–4, 2007.
springerlink.com

toward normal. Conversely, a decrease in blood glucose stimulates glucagon secretion; glucagon then functions in the opposite manner to increase the glucose concentration towards normal [4].

1.2.2 Historical Background

Before insulin was discovered, patients with diabetes suffered from polyuria and a catabolic state, which depleted them of strength, weight and fluid. As described by Arestaeus of Cappadocia in the 2nd century,

> "Diabetes is a dreadful affliction, not very frequent among men, being a melting down of the flesh and limbs into urine. The patients never stop making water and the flow is incessant, like the opening of aqueducts. Life is short, unpleasant and painful, thirst unquenchable, drinking excessive, and disproportionate to the large quantity of urine, for yet more urine is passed. One cannot stop them either from drinking or making water. If for a while they abstain from drinking, their mouth becomes parched and their body dry; the viscera seem scorched up, the patients are affected by nausea, restlessness and a burning thirst and within a short time, they expire." (Adapted from [5]).

For centuries, the only available treatment for patient with diabetes was starvation. If the diabetes worsened, then more starvation was prescribed.

The discovery of insulin at the University of Toronto in 1921–1922 was one of the most important milestones in the history of medicine [6]. The team that discovered insulin was quickly awarded a Nobel Prize in 1923.

1.2.3 Diabetic Patients

In patients with diabetes mellitus, the correlation between insulin delivery and blood glucose concentration is impaired. Both Type I and Type II diabetics have a dysfunctional endocrine pancreas, which produces little (as in Type II) or no insulin (as in Type I). Apart from this, the insulin receptor on the tissue cells in Type II diabetics also respond abnormally to the circulating insulin ("insulin resistance"). Type I diabetes is also known as Insulin-Dependent Diabetes Mellitus (IDDM), while Type II is also known as Non-Insulin-Dependent Diabetes Mellitus (NIDDM).

1.2.4 Surgical Patients

Surgical patients commonly enter a hypermetabolic stress state, induced by the area of infection or injury, and promoted by organs involved in the immunologic response to stress [7]. In this stressed state, insulin secretion is suppressed, and the normal carbohydrate metabolism is altered. This results in increased glucose production, depressed glycogenesis (i.e. reduced conversion of glucose into the storable form of glycogen), glucose intolerance, and insulin resistance [7].

1.3 Importance of Tighter BG Level Control

BG level should be kept within the normal range, because:

1. a high glucose concentration exerts an osmotic pressure in the extracellular fluid, and can cause cellular dehydration.

2. too low a BG level carries the risk of hypoglycaemic coma. Glucose is the only source of energy that can be used by the brain. Prolonged and profound hypoglycaemia can produce severe brain damage.

3. too high a glucose concentration (>11.1 mmol/l) can affect wound healing and interfere with human neutrophil function [8].

4. therapy that maintains BG level at below 11.9 mmol/l has been shown to improve the long-term outcome of diabetic patients with acute myocardial infarction [9].

Researchers in the last few decades have found that the mere use of insulin alone is not enough to guarantee the well-being of the patient (see e.g. [5, 10]), as diabetic microvascular complications have emerged. These microvascular complications include retinopathy (visual impairment), nephropathy (kidney disease) and neuropathy (nerve damage).

Studies in Type I [11] and Type II [12,13] diabetic patients have found that the onset and progression of these serious complications can be delayed by intensive insulin therapy. This intensive therapy includes the administration of insulin three or more times daily by injection or an external insulin pump, with the goal of achieving preprandial blood glucose concentrations of between 3.9 mmol/l and 6.7 mmol/l, postprandial concentration of less than 10 mmol/l, a weekly 3 a.m measurement greater than 3.6 mmol/l, and haemoglobin A_{1c} (glycosylated haemoglobin) within the normal range (less than 6.05%) measured monthly. The insulin dosage was adjusted according to the results of self-monitoring of blood glucose (SMBG) performed at least four times a day, dietary intake, and anticipated exercise.

The complications listed above occurred notably in diabetic patients, but also impacted on the critically ill patient population:

1. Findings from [9] has shown that maintaining BG at a level that did not exceed 110 mg/dl (6.1 mmol/l) substantially reduced mortality and morbidity in critically-ill patients in the ICU. In addition, a pronounced hyperglycaemia in critically-ill patients, even those who have not previously had diabetes, may lead to complications in such patients [9].

2. Patients with mean glucose concentrations >11.1 mmol/l within 36 h following surgery were more likely to develop infectious complications than their counterparts who were under better glycaemic control [14].

All these findings demonstrated that apart from the administration of exogenous insulin being essential to control the blood glucose concentration and to maintain homeostasis in patients, the tightness of such control is crucial in avoiding long-term complications (in diabetic patients) and infectious complications (in critically ill patient). The finding in [9] sets a desirable BG target for a closed-loop algorithm to achieve.

1.4 Achieving Tighter Control

In an effort to achieve tighter control, attempts have been made to sample the patients' BG level regularly, and adjust the insulin infusion rate automatically, so as to steer the BG towards a given target value (see e.g. [15], [16] etc). In such early experiments, whole blood was sampled continuously for glucose measurement, using the invasive method (see Section 2.2). Good glycaemic control was reported using this approach.

On the other hand, clinical routines have progressed with manual BG level control, which uses intermittent (and not continuous) BG level sampling. BG samples are taken from finger pricks or in-dwelling cannula, hourly, 2 hourly or 4 hourly, depending on severity. Various algorithms for perioperative control of BG level have been proposed (see Section 4.2), using hourly and three-hourly whole blood BG level sampling. Closed-loop (automatic) systems were not practical for use in routine treatment, due to the cost and preparation associated with their operation. Furthermore, long term use of the system posed safety concerns, due to the invasive nature of the measuring technique.

1.5 Conclusion

It is important to tightly control blood glucose level, both in ambulatory and hospitalised patients, due to the demonstrated benefits. However, achieving tight control requires frequent sampling of BG value, which to-date has been painful, difficult (and expensive). There are efforts in-progress to reduce or alleviate the pain of BG sampling, and also to improve the regulation of BG. These progress will be discussed in the chapters to follow.

Glucose Control: Input and Output

2.1 Introduction

Automatic regulation of a patient's blood glucose (BG) level requires a minimum of three components, namely, a continuous BG sensor, a controller that matches BG level with an appropriate insulin delivery rate, and an infusion pump to deliver the insulin to the subject.

Shown here is a simple model of the control loop:

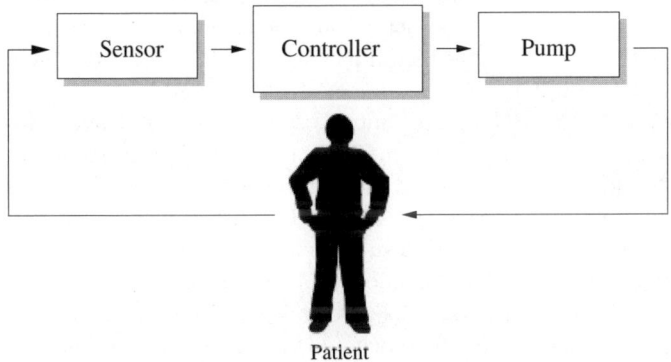

Patient

As the variable to be controlled is a patient's BG level, a knowledge of the BG is required. This is provided by a glucose sensor, and represents the input to the control system. Since insulin is used to lower a high BG level, the rate of insulin delivery represents the output of the control system. The patient is the "plant" to be controlled in control system terminology.

The controller is the component of the system that regulates the blood glucose levels in the patient. The formulation of the control rule depends on the knowledge we have about the sensor, the pump and the patient - and specifically, the BG measurement methods, the type (or preparation) of insulin used, the route of infusion, and the patient's characteristics. Various BG measurement techniques exist, and each has its unique characteristics. As for insulin,

F. Chee & T. Fernando: Closed-Loop Control of Blood Glucose, LNCIS 368, pp. 5–48, 2007.
springerlink.com   © Springer-Verlag Berlin Heidelberg 2007

each type of insulin has different kinetics, and different infusion routes exhibit different characteristics.

This chapter reviews:

- the various BG measurement techniques in existence to-date, which may generally be categorised into:
  1. Invasive,
  2. minimally invasive, and
  3. non-invasive techniques.
- the properties of the different types of insulin and their routes of infusion.

2.2 BG Measurement – Invasive Techniques

Historically, measurement of BG has been done via direct venous access, and subsequent ex-vivo determination of whole blood glucose levels by a glucose sensor (as used in [16,17]). As this was an invasive method, patients needed constant medical supervision [18,19], while a portion of their blood was channelled from a vein into a dual-lumen catheter, mixed with heparin (to prevent blood from clotting in the tubing), and sent to a measuring device [16,17]. The heparin does not enter the patient's circulation.

In the measuring device, the BG level was usually measured based on an immobilised glucose oxidase combined with amperometric detection of either peroxide, oxygen, changes in the pH or changes in colour [16,17]. Other detection methods include potentiometric detection, calorimetric detection, or a number of other variations of indicator techniques. More details of these techniques can be found in [20] and [21].

The latency of the BG measurement process (i.e. time taken from venous blood extraction to ultimate measurement) was initially 10 min, but was reduced to 1.5 min with an improved measurement technique. The promptness of measurement was important in the 1980s, because the so-called "artificial endocrine pancreas" regulated the insulin delivery on a minute-by-minute basis. It was reported that any delay of more than 6 min in the measurement would prevent effective feedback action [22]. Nowadays, on-line glucose analysers have a measurement latency of 50 sec (such as the Via Blood Glucose Monitor from Metracor, San Diego, CA).

Advantages

- The advantage of this method is the accuracy of the measurement, as whole blood is analysed for its glucose content, i.e. the measurement is made in the actual fluid of interest.

Disadvantages

- Since invasive BG measurements continuously withdraw blood for analysis, any studies conducted using the technique are limited in duration as described below. The withdrawn blood is usually discarded (as it is mixed with heparin, and is not safe to be returned to the patient).

• The practicability of long-term invasive measurement for closed-loop glucose control is mainly hindered by concerns for patient safety, because this technique carries the risk of infection and thrombosis. Furthermore, the fact that the technique cannot be performed by patients themselves without proper supervision [19, 23] limits the technique to be used only in hospitals.

2.3 BG Measurement – Minimally-Invasive Techniques

Minimally invasive methods provide a more favourable solution than invasive methods, mainly because the measurement of glucose is now done outside the vascular tree, minimizing the risk associated with vessel access (i.e. pain, infection and thrombosis).

These methods measure the blood glucose level indirectly — using the glucose level in the interstitial fluid that surrounds all cells under the skin (or subcutaneous space) to reflect the plasma glucose level. This technique is based on the assumption that a correlation exists between the subcutaneous (s.c.) glucose level and the plasma (or whole blood) glucose level.

This assumption is valid to an extent, as many studies have reported on the relationship of subcutaneous glucose to blood glucose [24, 25]:

• At equilibrium, the glucose concentration at s.c. sites correlated well with those of plasma [25, 26].
• During periods when the blood glucose concentration is rising or falling, the two concentrations differ [27] with a delay of an average of 10 min in the s.c. tissue [28]. But delays as much as 45 min have also been reported [25].

Sites such as adipose tissue and s.c. tissues are regarded as most appropriate for minimally-invasive BG determination because of the ease of accessibility for surgery or sensor replacement [29]. Also, less pain receptors exist at these sites [24].

Measurement of s.c. glucose can be done by direct implantation of a glucose sensor in the s.c. tissue, or by means of fluid extraction from the s.c. space (usually by microdialysis), as discussed below.

2.3.1 Implantation

In "implantation", miniaturised glucose sensors were inserted directly into the s.c. tissues [30, 31], and measurement was performed in situ. Once the sensor is inserted, glucose level can be determined using methods like:

• amperometry
• fluorescence detection

Amperometric Glucose Sensor

Among the chemically-based methods of glucose sensing, the amperometric sensor is one of the most simple, stable, and widely studied (and perhaps the most successful) approaches [21]. As described in [32],

> "amperometry is the determination of the intensity of the current crossing an electrochemical cell under an imposed potential. This intensity is a function of the concentration of the electrochemically active species in the sample."

Amperometric glucose sensors are almost entirely based on the glucose-oxidase (GOx) enzyme, with the flavin-adenin-dinucleotide (FAD) radical as the active part [33]. FAD is able to react with hydrogen to form $FADH_2$. The two forms of GOx are thus marked as "GO(FAD)" and "GO($FADH_2$)". The chemical principles underpinning amperometric methods of measuring glucose concentration are described in Table 2.1.

Table 2.1. Glucose-oxidase reaction

Glucose-Oxidase Reaction

Glucose oxidase dehydrogenates glucose ($C_6H_{12}O_6$), transforming it to gluconolactone ($C_6H_{10}O_6$):

$$Glucose + GO(FAD) \longrightarrow gluconolactone + GO(FADH_2)$$

And:

1. Gluconolactone reacts with water to form gluconic acid ($C_6H_{12}O_7$), which dissociates in solution.

$$Gluconolactone + H_2O \longrightarrow C_6H_{12}O_7 \Longleftrightarrow C_6H_{11}O_7^- + H^+$$

2. Glucose oxidase is then oxidised in the presence of oxygen, and hydrogen peroxide, H_2O_2, is formed as a result.

$$GO(FADH_2) + O_2 \longrightarrow GO(FAD) + H_2O_2$$

By applying -600 mV to -700 mV (with respect to the working platinum electrode), H_2O_2 will be electrolysed into O_2 and $2H^+$ and electrons flow as a result.

$$H_2O_2 \longrightarrow O_2 + 2H^+ + 2e^-$$

This generated current is linearly dependent upon the concentration of glucose.

Since oxygen is a limiting factor in the reaction, other methods that use mediator to eliminate the dependence on oxygen also exist. The reaction described above is only representative.

When amperometric sensors are used as implantable sensors, enzyme immobilisation techniques such as cross-linking of glucose oxidase and Bovine Serum Albumin (BSA) to the electrodes using glutaraldehyde (GA), are frequently used to ensure maximal contact and response. Immobilisation has the advantage of stabilising the protein so that it can be used repeatedly [32]. The use of BSA improves enzymatic activity because of the better mass distribution of the various proteins. This configuration (i.e. GOx + BSA + GA) has a response time of 10 sec [32], and this is the reason behind the CGMS sensor taking glucose readings every 10 sec (because it uses the same configuration (see [34])).

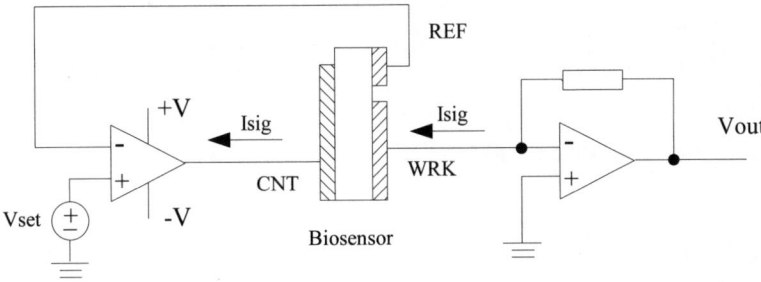

Fig. 2.1. Basic circuit illustrating the operation of amperometric sensors (Adapted from [33])

Implantable amperometric sensors have only recently become available as an off-the-shelf product. Medtronic MiniMed Inc (Northridge, CA) is the first company to market a FDA-approved amperometric glucose sensor for use in human subjects. Called MiniMed CGMS (Continuous Glucose Monitoring System), it is designed for continuous BG monitoring in diabetic patient for periods of up to 3 days. CGMS takes a subcutaneous glucose reading every 10 sec and stores an averaged reading every 5 min. The sensor is pre-sterilised, readily-implantable, easily inserted, and features a short (1-hour) preparation time. The short preparation time is an advantage over the 2-hour run-in period required for other earlier attempts [35].

Fluorescence Detection

Fluorescence-based glucose sensor involves measuring a change in light emission when glucose binds to certain molecules (in affinity-binding process) or reacts with certain molecules (in glucose-oxidase reaction).

- Affinity-binding glucose sensors uses competitive binding between glucose and a suitably labelled fluorescent compound, to a common (and glucose-specific) receptor site [38]. These fluorescence chemicals are immobilised within a selectively permeable membrane at the tip of an optical fibre sensor probe.

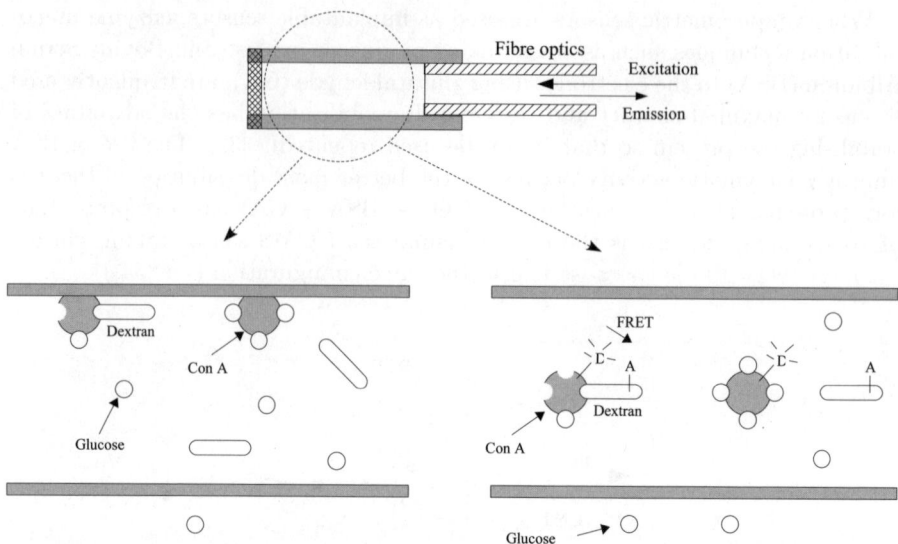

Fig. 2.2. [Left] Con A is immobilised on the inside of a hollow dialysis fibre. A low molecular weight cutoff of the dialysis fibre retains the dextran within the fiber lumen while allowing glucose to pass freely through the dialysis membrane [36]. Excitation light passes through the fibre optics and into the solution, fluorescing the unbound FITC-dextran. The resulting small fluorescence intensities are measured through the same fibre optics by a measuring system [37]. [Right] FRET mechanism allowing an acceptor-labelled dextran to quench the fluorescence when a donor is nearby (see text).

1. Schultz et al [36] used a protein called Concanavalin A (or Con A)[1] as the common receptor, with a fluorescein-isothiocynate labelled dextran[2] as the competing ligand. An increased presence of glucose will displace with the dextran from Con A, leading to an increase in fluorescence intensity (Fig 2.2).

2. In more recent work (e.g. McCartney et al [39]), the phenomenon of Fluorescence Resonance Energy Transfer (FRET) was used (Fig 2.2). In this method, Con A is labelled with a fluorophore donor, and the dextran is labelled with an acceptor. When the dextran binds to Con A, the donor is closer to the acceptor. As the donor is excited at its specific fluorescence excitation wavelength, this excited state is nonradiatively transferred to the acceptor (FRET phenomenon), resulting in a diminished fluorescence. Conversely, the displacement of dextran by glucose reduces FRET and increases fluorescence intensity and lifetime [41].

---

[1] Con A is a plant lecitin [39], extracted from jack bean [40].

[2] Dextran is a carbohydrate derivative [40], and is made of many glucose molecules joined into chains of varying length.

3. Ballerstadt et al [42] has proposed subcutaneous implantation of a self-contained semipermeable membrane chamber containing the glucose-sensitive fluorescent reagents close to the skin surface. Glucose measurement can be performed by illuminating the implant region. Similar idea has been trialled and presented by PreciSense A/S (Horsholm, Denmark) as part of the company's goal to develop a stable and precise minimally-invasive commercial FRET-based glucose sensor with 2 weeks continuous use (See [43]).

4. March et al [44] describe the incorporation of fluorescent-based glucose sensor into intra-ocular lens. The lens contains tetramethylrhodamine isothiocyanate Concanavalin A (TRITC-Con A) and FITC-dextran. When TRITC-Con A binds competitively with FITC-dextran and glucose, the bound FITC-dextran fluoresence is quenched through FRET. An increase in glucose concentration will liberate FITC-dextran, producing a fluorescence proportional to the glucose concentration.

- Glucose oxidase catalytic reaction can be monitored by detecting oxygen consumption or hydrogen peroxide production via fluorescent (see the reaction equations in Table 2.1). Glucose concentration can be quantified:

1. using fluorescence that can be quenched by oxygen [40], or fluorophore that is sensitive to oxygen concentration [38,45];
2. using redox mediator dye [46], where the dye (instead of water) reacts with gluconolactone to produce current;
3. detecting the formation of fluorescent oxidation product by hydrogen peroxide [38,47].

There are other fluorescence and affinity-binding methods in existence. Interested readers in the subject are encouraged to read [40].

Advantages of Implantable Sensors

- Implantable sensors are small in size, portable and do not require extraction of fluids for operation.

Disadvantages of Implantable Sensors

- The most common problem inhibiting the longevity and reliability of implantable sensors is membrane biofouling. Biofouling is the adhesion of proteins and other biological matter on the sensor surface, and in some cases, impregnation of the material [48], resulting in drifting and/or diminishing of the sensor signal. Biofouling starts immediately upon contact of the sensor with the body. Biofouling remains a problem in both amperometric and fibre-optic based glucose sensor [49].
- Biological matter including the immune cells can attack the glucose sensor assembly (especially the membrane) [50,51], causing inflammation.

- In the case of amperometric sensor, the body's warm and saline environment can corrode the implanted metal electrodes and inactivate the enzymes.
- A person's movement can create artifacts and noise in the measurement.
- Substances such as vitamin C and acetaminophen may react at the electrode, depleting the hydrogen peroxide before it can react at the electrodes [51]. Other problems include the oxidation of other reactants at the same potential as glucose oxidase, creating noise and inaccuracy in the measurement.
- Although Con A and glucose react very rapidly with time constants in the order of milliseconds [42], the response time of the affinity sensor was reported to be 7 ± 2.5 min [52]. The response time are limited by the diffusion rate of glucose into or out of the sensor, the diffusion of FITC-dextran within the hollow fibre lumen, and the temperature.

2.3.2 Fluid Extraction

Some existing fluid extraction techniques include:

- microdialysis;
- transcutaneous harvesting by reverse iontophoresis;
- transdermal extraction by ultrasound;
- microporation with a laser beam (and subsequent fluid extraction).

Microdialysis

Microdialysis is a method that allows blood or interstitial fluid (ISF) to be sampled and analysed by an ex-vivo sensor. Hollow fibres of dialysis membrane are implanted, often subcutaneously, and perfused at low flow rate with isotonic fluid (Fig 2.3). The dialysate is pumped to a flow chamber that incorporates a glucose sensor [53].

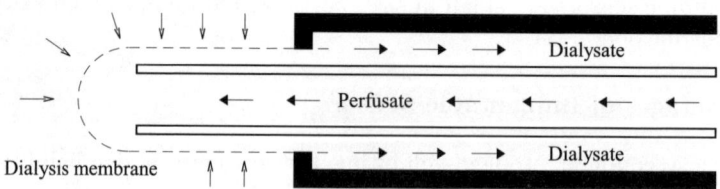

Fig. 2.3. The inner cannula of the dialysis probe is perfused with an isotonic fluid (perfusate), which is transported to the tip of the dialysis probe. The probe is placed under the skin, and substances (e.g. glucose) having a concentration gradient compared with the perfusate, diffuse through the dialysis membrane down the concentration gradient (Adapted from Meyerhoff et al [26]).

Advantages with Microdialysis

- Microdialysis methods prevent many of the other substances present in the ISF from entering the measuring cell and creating noise in the measurement.

Disadvantages with Microdialysis

- Although microdialysis filters off much of the unwanted substances preventing them from entering the measurement chamber, the catheters used in microdialysis probes are very fragile and difficult to use in the clinical environment. In addition, biofouling will still happen at the "exchange" membrane of the dialysis probe (rather than at the glucose sensor in the case of implanted sensor).

- The use of microdialysis systems requires a method to dispose off the dialysate, which may pose a health risk if not done properly (e.g. due to the potential for viral transmission such as hepatitis B and C and HIV).

Reverse Iontophoresis

Reverse iontophoresis involves the application (or conduction) of a constant, low-level electrical current to the skin between an anode and a cathode. The applied potential draws sodium and chloride ions from beneath the skin toward the cathode and anode, respectively. Glucose is also carried along with these ions by electro-osmosis, and transported across the skin. Since the skin has a negative charge at neutral pH, there is a greater net transport to the cathode. As a result, glucose is preferentially extracted at the cathode [54–56]. This transdermally extracted glucose is then measured by a standard amperometric technique involving glucose oxidase.

Although the glucose concentration of the extracted fluid is minute (about ~1/1000 that of blood glucose) [54], it is shown to be proportional to that of plasma blood [55–57]. This technique is currently being marketed by Cygnus Inc (Redwood City, CA) as GlucoWatch G2 Biographer, offering near non-invasive measurement. See Section 2.5.3 for a technical description of GlucoWatch Biographer.

Advantages with Reverse Iontophoresis

- This technique is near non-invasive. Despite reports of mild discomfort when using the device [51, 55], this technique is relatively painless compared to implantable sensors.

Disadvantages with Reverse Iontophoresis

- Because of the use of a low current to extract glucose, it takes 10–20 min from the beginning of each fluid extraction cycle until a blood glucose level can be reported [56]. Since extracted glucose comes from the subcutaneous space, there is a chance that any existing delay in glucose diffusion between the subcutaneous space and the plasma will be added to the fluid extraction cycle delay. There is potential for a long lag phase in the detection of blood glucose changes.

- Excessive sweating can prevent accurate measurement, as the device depends on minute glucose concentrations in ISF, which may be contaminated by sweat [56].

Ultrasonic Transdermal Extraction

Ultrasonic transdermal extraction uses ultrasound to facilitate the transport of glucose from the ISF across the skin for non-invasive monitoring. It was reported that sufficient amount of clinically relevant analytes including glucose can be extracted using low-frequency ultrasound. Ultrasound is defined as sound that has frequency beyond 20kHz. The mechanism of improved transdermal transport by ultrasound is still not well understood [58]. It is suggested that ultrasound can induce cavitation in and around the skin. Oscillation and collapse of cavitation bubbles disorder the lipid bilayers of the skin, resulting in enhanced skin permeability [59]. It is also reported that once treated with ultrasound, the skin permeability remains high for up to 15 hours [58].

Suggested configuration of the glucose monitor would feature an ultrasonic device to permeabilize the skin (for ∼2 min) and a patch to be placed on the ultrasonically permeabilized skin for continuous extraction and detection of glucose [60]. Sontra Medical Corporation (Franklin, MA, USA) is in the process of developing an ultrasonic glucose monitor called "Symphony™ Diabetes Management System" (see http://www.sontra.com).

Advantages with Ultrasound

- This technique is near-invasive, with no pain and no visible effect of ultrasound on skin.

Disadvantages with Ultrasound

- A contact medium (e.g. oil, water oil emulsions, ointments) between the ultrasonic probe and skin is required. Otherwise, the ultrasound will be completely reflected by air [60].
- A lag time exists between capillary blood glucose concentration and the extracted glucose concentration (due to diffusional delay) [60].
- Although no adverse effects of ultrasound was reported, further research focusing on safety issues is required [58].

Microporation

Microporation involves puncturing the skin layer using a micro laser beam. The laser creates a micro-hole on the surface of the skin, and glucose is extracted by means of suction (vaccum) or a patch [61]. The extracted glucose is then measured by a conventional method (such as amperometry). Products incorporating this method have been described by companies, such as SpectRx Inc (Norcross, GA) (http://www.spectrx.com) in [62], and PowderChek Diagnostics (Fremont, CA).

2.4 BG Measurement – Non-invasive Techniques

Without argument, the most welcomed BG measurement would be a painless and "finger-prick free" measurement. Such a measurement would require a means of examining the glucose content in the tissue without puncturing the skin, or causing discomfort (such as skin irritation etc). The non-invasive techniques include:

1. Optical spectroscopy;

    Glucose can be measured by optical means because it exhibits optical properties such as light absorption, reflection, polarisation, circular dichroism and other responses to radiation. Methods that allow the optical properties of glucose to be identified can be used to determine glucose concentration non-invasively. Most measurement techniques used in analytical chemistry have been used to measure glucose.

2. Dielectric spectroscopy (or impedance spectroscopy).

At the time of writing, none of the reported non-invasive technology has been used in routine clinical practice [58], and progress are being made to bring these non-invasive system to commercialisation.

2.4.1 Infrared Absorptiometry

One of the most studied methods for glucose determination by its optical properties is the infrared absorptiometry. Infrared absorptiometry measures the absorption of light intensity by a solute (e.g. glucose) as light passes through a solution. By determining the light attenuation caused by absorption at a single wavelength [63], the glucose concentration in a solution can be quantified. However, to obtain an accurate reading, the glucose solution has to be clear because light scattering would result in additional attenuation of light.

In a complex mixture of substances (such as tissue), it is still possible to quantify glucose by using various wavelengths and complex mathematical procedure like multivariate calibration. The more the spectra of the substances are different from each other, the better the reliability of the quantification [63]. The ex vivo measurement of glucose in plasma, serum or whole blood (i.e. complex structures) is feasible using high performance equipment and sophisticated mathematical calibration procedures [63, 64].

Optical absorption technique are based on selective absorption of light by the molecule. The absorption can be described by the Beer-Lambert law [38]:

$$I = I_o e^{-\epsilon C L}$$

where I_0 is the intensity of incident optical radiation, I is the transmitted intensity, ϵ is the wavelength-dependent molar extinction coefficient[3] $[(\text{mol/l})^{-1}\text{cm}^{-1}]$, C is the molar concentration, and L is the path length [cm].

The absorbance can be defined by

$$A = \log_{10} \frac{I_0}{I}$$

Infrared absorptiometry works because each molecule has specific resonance absorption peaks [66], i.e. the specific frequencies at which they vibrate corresponding to the excitation source. In another word, the frequency of the absorbed light is the molecular vibrational[4] frequency actually responsible for the absorption process [67]. Each organic molecule has an unique chemical composition and thus an unique infrared spectrum. Optical measurement of glucose can be performed using Near-infrared or Mid-infrared radiation.

Near-IR [700 nm–2.5 μm or 14000–4000 cm^{-1}]

The near-infrared (Near-IR) spectrum extends from the end of the visible spectrum[5] to the start of the fundamental infrared absorption bands at 2500 nm. Absorption in the Near-IR region are most often associated with overtone and combination bands of the fundamental molecular vibrations of C–H, O–H and N–H functional groups [38, 64, 65].

Near-IR light can penetrate deeper into the tissue (than other infrared frequency bands). In addition, within the spectral range of 600–1300 nm [41], there is an "optical window" where tissues are "transparent" to light.

Glucose produces one of the weakest Near-IR absorption signals per concentration unit. In addition, the normal proportion of glucose in blood and tissue is only about 0.1% of the water content [68]. Measurements of glucose in tissues by optical means are thus limited by the signal-to-noise ratio that can be achieved at such low concentrations. Furthermore, measurements can be confounded by issues such as

- interference by water, which absorbs very strongly in the Near-IR region;

  Water exhibits very similar absorption spectra to glucose [63], and absorbs very strongly in Near-IR region [64]. Glucose absorption peaks has small magnitude compared to a large aqueous background spectrum, and this yields low signal-to-noise ratio.

---

[3] Also known as Molar absorptivity, molar extinction coefficient is the absorbance A of a solution divided by the product of the optical path l [cm] and the molar concentration C of the absorbing species [65], i.e. $\epsilon = A/(l \times C)$.

[4] There are many types of molecular vibrations. Readers are invited to consult general textbooks on Analytical Chemistry for more details.

[5] Visible spectrum ranges from 380 nm (violet) to 700 nm (red) approximately, with no clear boundaries between colours.

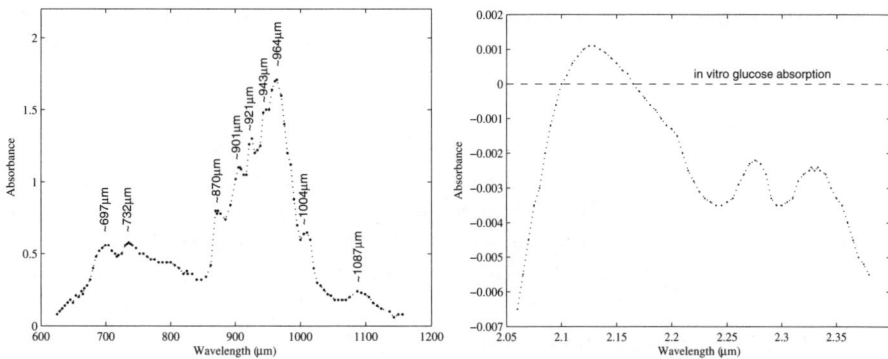

Fig. 2.4. Glucose absorption spectrum (Adapted from McNichols et al [38])

- strong light scattering by skin [63], as well as other substances and compounds in tissues (such as blood cells) [41];

- overlaps in light absorption by molecules introduces interference into the absorbance measurement;

  Since biological molecules have complicated structure, they exhibit a large number of similar IR absorption peaks that are often overlapping [66] (e.g. glucose has more than 20 absorption peaks in the wavelength region of 2.5–10 μm). Not all of these peaks are specific for glucose [63]. The prominent absorption peak of glucose is around 9.7 μm. In addition, haemoglobin is the most dominant component in blood, with a concentration level of more than 100 times of glucose. Its absorbance becomes increasingly strong toward short Near-IR and visible wavelengths. Although haemoglobin absorption peaks do not interfere with those of other components, its influence is by no means negligible due to its high concentration [69].

- pronounced temperature dependence of light absorption [63, 69].

Because of the strong absorbance of water, Near-IR transmittance measurement should use regions between the water bands where sufficient amount of light can be transmitted [70]:

- Near-IR region between 0.7–1.3 μm — This range contains higher orders of glucose overtone regions [70]. Nevertheless, this region also shows very little glucose absorption (i.e. less than 0.1% of the fundamental region of glucose at 9–9.6 μm) despite having lower light absorption by water [69].

- Near-IR region between 1.5–2.5 μm — This range was reported to be a suitable for noninvasive glucose monitoring [69], with highest glucose absorption at 1.575–1.7 μm (which corresponds to the overtone band of C-H stretching mode [71]). Most of the analytes of interest display distinguishing absorption features that do not suffer from excessive attenuation by water [72].

- Near-IR region between 2.0–2.5 μm –This range is also popular for aqueous glucose measurements because it contains a relative minimum in the water absorption spectrum and readily identifiable glucose peak information.

With all the confounding issues, extra signal processing (in conjunction with Near-IR measurement) became necessary, e.g. digital filtering to correct for temperature sensitivity of the Near-IR spectrum, and multivariate calibration techniques. Sensing site for Near-IR measurement included fingertip [41], earlobe, tongue, lip [73], forearm, oral mucosa surface and aqueous humour [38].

Mid-IR [2.5–50 μm or 4000–200 cm^{-1}]

Mid-IR spectrum ranges from 2.5–50 μm [67], and this region is often called "fingerprint" region. This is because the absorption bands in this region are due to the fundamental stretching and bending modes of the molecule. Mid-IR spectrum is very useful for spectral identification of compounds [38].

The Mid-IR range of 5–13 μm contains absorbance fingerprints generated by biologically important molecules such as glucose, proteins and water [74]. The prominent absorption peak of glucose is reported to be around 9.7 μm [75,76].

However, water absorbs Mid-IR radiation very effectively [63,76]. Background absorption bands of other solution constituents are also strong, and thus limits the depth and path length of Mid-IR transmission spectroscopy to a few hundred microns. [38,63,77]. Alternative measurement techniques using Mid-IR radiation have been proposed (see Section 2.4.2).

Implementation

The basic setup of the IR measuring instrument consists of a light source, an optical assembly (that directs the light at a sample and collects the transmitted or reflected light), and a light detector unit (which contains a spectrometer if measuring over multiple wavelength). The measured spectra were then analysed using mathematical procedures, most commonly the Multivariate Calibration methods like PLS (Partial Least Square) and PCR (Principal Component Regression), or a Neural network [78]. Multivariate calibration (in the simplest form) involves multiplying the (measured and processed) spectrum A_i by a vector of calibration coefficients b_i, and summing the result over all wavelengths to obtain the analytes concentration c [72]:

$$c = \sum_i A_i b_i + \text{constant}$$

Some examples of IR absorptiometry instrumentation (and not meant to be exhaustive),

- Robinson et al [79] used an FTIR-spectrometer with a tungsten halogen source and a liquid-nitrogen cooled InSb (indium antimonide) detector (or a gating gating spectrometer, tungsten halogen source and Si array detector), over the frequency range 870–1300 nm. PLS (Partial least square) and PCR (principal component regression) calibration were applied to define an empirical model relating changes in the calibration spectra to the differing reference glucose concentrations of blood samples obtained during the calibration. Burmeister et al [70] and Small et al [80] also used FTIR spectrometer with a tungsten-halogen lamp to collect the Near-IR spectra.

- Hiese et al [81] used attenuated total reflection (ATR)[6] and FTIR spectroscopy to estimate the blood glucose concentration in whole blood sample (taken out of subjects) over the wavelength range 1500–750 cm^{-1} (or ~660–1300 nm). Multivariable calibration with the PLS algorithm was performed on the spectral data.

- Tenhunen et al [82] used a fibre bundle as a transflectance probe to direct the light from a halogen lamp to achieve an appropriate penetration depth into the tissue. Temperature stabilised IR sensitive detector was used for measuring the interferogram of the FTIR-spectrometer.

- Yoon et al [71] did not use a spectrometer, but used custom made detectors to measure the absorbance of the Near-IR light. A minimum of three different wavelengths were selected for determination of glucose (λ_1=1625 nm, λ_2=1365 nm, λ_3=1200 nm), based on satisfying a simple but efficient algorithm that will estimate the glucose concentration. Instruments were setup up carefully to minimise reflected light from skin surface and effects of tissue scattering.

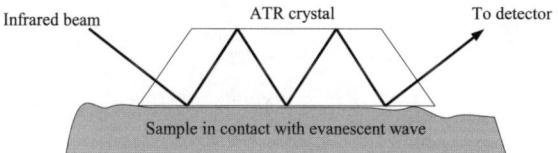

Fig. 2.5. Attenuated Total Reflection: When an infrared beam is directed onto an optically dense crystal with a high refractive index at a certain angle, the internal reflectance creates an evanescent wave that extends a few micrometer (0.5–5 μm) beyond the crystal surface into the sample held in contact with the crystal. As the sample absorbs the relevant infrared spectrum, the evanescent wave will be attenuated. The attenuated energy from each evanescent wave then passes back to the IR beam, which then exits the opposite end of the crystal and is passed to the detector in the IR spectrometer (Adapted from PerkinElmer Inc [83]).

---

[6] ATR is a sampling technique that allows samples to be examined by a spectrometer directly in solid or liquid state without extra preparation. It measures the changes that occur in a totally internally-reflected infrared beam when the beam comes into contact with a sample. Kaiser [73] was among the earliest teams to combine ATR prism with laser absorption spectrometry for measurement of glucose on the lips.

Table 2.2. FTIR-spectrometer

Fourier-Transform Infrared Spectrometer (FTIR)

Spectrometer (or spectroscope) is an optical instrument that disperses light into its spectrum, allowing measurement of the individual wavelengths and intensities.

In the original spectrometer design, light entered a slit and a collimating lens transformed the light into a thin beam of parallel rays. A prism then separated the beam into its spectrum. The observer then viewed the spectrum through a tube.

Later design used diffraction gratings in place of the prism. Diffraction gratings have fine parallel and equally spaced grooves on the surface. Each wavelength of the incident beam spectrum is sent into a different direction when the gratings is illuminated by light. (Every day examples of gratings are CD and DVD media).

A Fourier-Transform Infrared (FTIR) spectrometer uses interferometric method to measure the spectrum, in contrast to the classical spectrometers. In its simplest form, an FTIR spectrometer consists of an infrared light source, a beam splitter, a fixed mirror, a moving scanning mirror and an infrared detector. This is the typical Michelson Interferometer design.

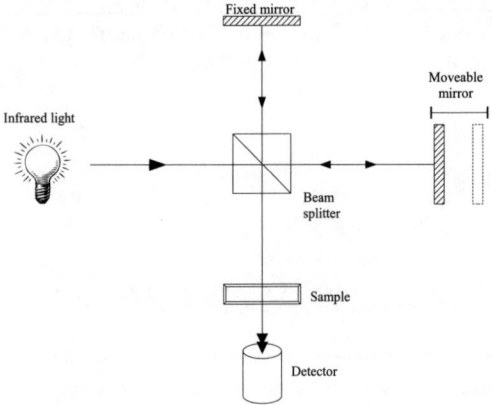

The IR beam is guided through the interferometer. After passing through the sample, the amplitudes of the waves are combined either constructively or destructively to form an interferogram (i.e. a plot of the output power from the detector as function of the difference in path-length of the two beams [67]). FTIR measured all the wavelengths simultaneously with a single detector element [65, 72]. By performing a mathematical Fourier Transform on this signal, a spectrum identical to that from conventional (dispersive) infrared spectroscopy can be obtained.

Different types of IR detector can be used. One common choice is indium gallium arsenide (InGaAs) detector element because it has good specific detectivity and excellent amplitude linearity over a large dynamic range required for interferometric signals [72]. It also cost less than the array detectors used in gating spectrometers.

It should be noted that as the IR wavelength is lengthened, the usable optical path length shortens because of more scattering and absorption in the tissue [72]. Reflection measurements, or back-scattering become more effective than transmission measurements in this scenario (see Section 2.4.3). Readers interested in further details on infrared glucose measurements are encouraged to read articles such as [84] and [85].

2.4.2 Mid-IR Emission Spectrometry

In contrast to infrared absorptiometry, the Mid-IR emission spectroscopy is based on the principle that the human body naturally emits strong electromagnetic radiation (or blackbody radiation), and the natural Mid-IR emission from the human body is modulated by the state of the emitting tissue [86]. The light source in this technology is the human body's natural emission of Mid-IR light (no external light source needed). The measurement techniques reported in the literature are:

- Temperature gradient (Cooling the surface [74,87,88])

  In the Mid-IR range, glucose absorbs strongly with minimal interference from other species [87]. However, glucose is not measureable at the same temperature as the tissue that emit the blackbody radiation [74,87]. To solve this problem, a temperature gradient is introduced, where the surface is cooled so that the cooler glucose will absorb more than it will emit. In another words, when there is a temperature gradient, the radiation intensity (which is wavelength-dependent) will deviate from the original Planck's prediction (of blackbody emission). At the wavelengths where glucose absorbs, the total energy emitted will fall below the Planck curve [74]. This would yield a reasonable glucose absorption spectrum superimposed on the usual smooth blackbody spectrum [87].

- Tympanic membrane measurement [86]

  Malchoff et al described the detection of glucose from the tympanic membrane using room temperature detectors and a specially designed non-dispersive filter-based spectrometer. An IR wave-guide with gold-plated inner surface directs IR radiation into a circular variable IR filter in close proximity. The circular variable IR filter has transmission band of 7.7–14.1 μm, and when rotated in the IR light path would generate variable passbands. A thermopile detector was placed very close to the filter surface to capture the light output from the filter. The detector window has two more filters – one that passes thermal emission bands with glucose signature; the other passes radiation that does not include emission bands characteristics of glucose at wavelengths in the range of interest.

2.4.3 Scatter Change

Light scattering is the next major optical interaction with tissue after light absorption [89]. Light scattering occurs when light interacts with variations in the refractive index[7] or small particles acting as scattering centres in the medium, resulting in light dispersing from a straight trajectory. When scattering centers are grouped together, the light may get scattered many times (i.e. multiple scattering).

In tissue, most of the light scattering can be attributed to Mie scattering, which is related to:

- the the size of the scattering particles;

  Mie scattering takes place when the size of the scattering particles and the wavelength of light are in the same order of magnitude [63].

- the magnitude of the refractive index mismatch between the particles and their medium [90].

  The amount of the scattering depends on the ratio between the refractive index of the scattering particles (e.g. cell membranes) and the surrounding medium (e.g. extracellular fluids). If the difference between the refractive indices is pronounced, light scattering is larger. With identical refractive indices, the media is transparent [63, 91].

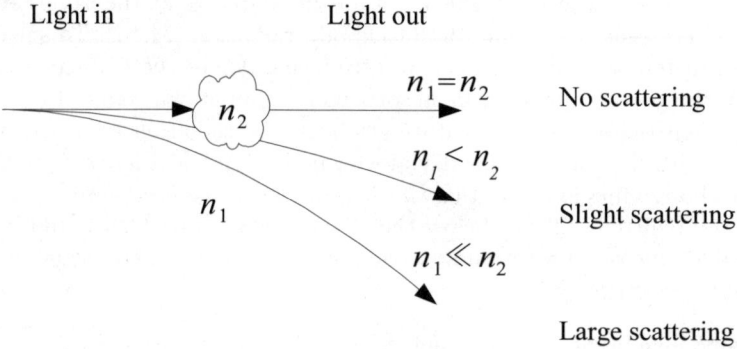

Fig. 2.6. Scattering of light (Adapted from Heinemann et al [63])

Due to the free exchange of glucose between blood plasma and interstitial fluid, changes in glucose concentrations of these two fluids are closely correlated. An increase in glucose concentration leads to a rise of the refractive indices of blood plasma and ISF (as glucose can displace the water from the fluids [92]), whereas the refractive index of the scattering particles (red blood cell, cell membranes, collagen fibres etc) is assumed to remain relatively unchanged [63, 91, 93]. Since the refractive index of ISF is lower than that of the scattering particles, an increase in glucose concentration causes a reduction of the refractive

---

[7] The refractive index (or index of refraction) of a medium is a measure for how much the speed of light is reduced relative to vaccum when the light travels inside the medium. For example, typical glass has a refractive index of 1.5, which means that light travels at $1 / 1.5 = 0.67$ times the speed in air or vacuum [65].

index mismatch between ISF and scattering particles. The increase in blood glucose concentration effectively decreases the scattering coefficient of skin [63]. The scattering coefficient is a factor that expresses the attenuation caused by scattering:

$$\mu_s = f\left(\rho,\ a,\ g,\ \frac{n_{cell}}{c_{medium}}\right)$$

where ρ = number of density of scattering cells in the observation volume, a = diameter of the cells, g = anisotropy factor (the average cosine of the angle at which a photon is scattered), n_{cell} = the refractive index of the cells, and n_{medium} = the refractive index of ISF [94].

It should be noted that the scattering coefficient of the skin is not directly influenced by glucose, but indirectly via the refractive indices [63]. The glucose effect on scattering coefficient is not specific to glucose molecule only. The glucose concentration measured from scatter change detection is derived from changes in refractive index "induced by glucose". Other blood analytes and physiological factors may influence the scattering coefficient, for example:

- mitochondria, which is the main scatterers in the skin [95];
- temperature [63, 92];
- changes in the intracellular refractive index;
- changes in cell size due to by osmolarity changes [91].

Implementation

To measure the scattering coefficient, a diffuse reflectance measurement is usually used. A narrow beam of light is directed through the skin surface into the tissue. Some of the light is absorbed by the tissue while a certain portion of the light is diffusely scattered and reflected back to the skin surface (after interacting with the tissue, or exciting vibrational modes of the analyte molecule in the tissue). The intensity of this diffuse reflectance is dependent on both the scattering coefficient and the absorption coefficient of the tissue.

"By measuring the reflected intensities at different distances from the light entry point, an intensity profile (i.e. diffuse reflectance vs. distance) can be recorded. The reflectance measured close to the light entry point is mainly influenced by the scattering of the skin, while reflectance further from the light source is affected by both the absorption and the scattering properties of the skin." (Adapted from Heinemann et al [89]).

The effects of light scattering and light absorption from the recorded light intensity can be differentiated by algorithms designed based on diffusion theory of light propagation in tissue [63]. A neural-network can be used to extract the optical properties from the reflectance data [91].

Examples of the measuring instruments include diode-array spectrometer with InGaAs detector [78], FTIR spectrometer with nitrogen-cooled InSb detectors [96], LED arrays with photodetector arrays [63], and He-Ne laser beam with CCD camera [97].

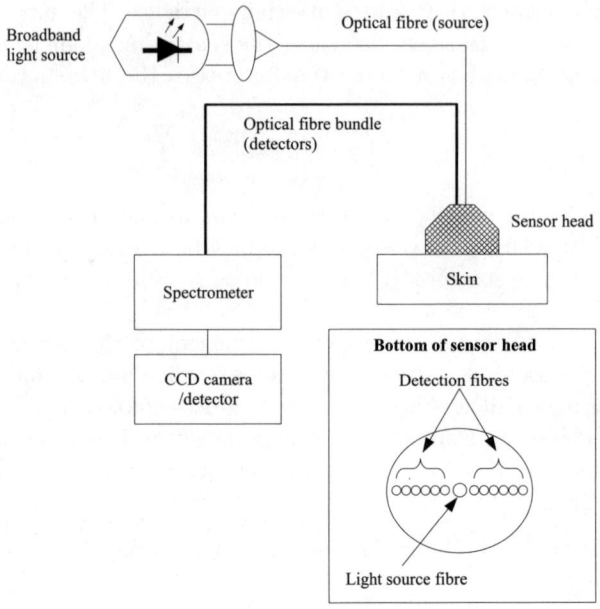

Fig. 2.7. Diffused reflectance for measuring scatter change (Adapted from Heinemann et al [63])

Optical Coherence Tomography has also been used to measure glucose based on the fact that OCT is very sensitive to the detection of changes in refractive indices of the sample, and thus the scattering properties of the sample. A reduction in the sample scattering property lowers the intensity of the backscattered photon. Larin et al [98] used OCT techniques to scan the skin up to a depth of 1 mm for measurement of tissue scattering properties at a specific layer in the skin. However, since OCT measures only relative changes in the scattering properties, other osmolytes of the ISF that can change the refractive index mismatch have the potential to interfere with glucose measurement [95].

2.4.4 Tissue Manipulation and Red/Near-IR Absorptiometry ($\lambda \sim 630 - 1300nm$)

In most Near-IR absorptiometry techniques considered above, optical non-invasive measurements were conducted using "DC measurement technique", where light illuminates blood-perfused tissue, and the reflection/transmission effect is studied. All spectrally active components of the tissues (e.g. skin, blood, muscle, fat etc) contribute to the observed effect. Orsense Inc (Nes Ziona, Isreal; http://www.orsense.com) in [101] considered an "AC measurement technique", which focuses only on the "blood" component of the tissue, by (for example) looking at the difference in light absorption of the tissue measured during the systole (blood pressure rise) and the diastole (blood pressure fall) period.

Table 2.3. Time-domain Optical Coherence Tomography (OCT)

Optical Coherence Tomography

Time-domain OCT is based on low-coherence interferometry (i.e. interferometer with a low coherence, broad bandwidth light source). It consists of a low-coherent light source, a detector, a reference arm and and a sample arm. The light is split into the reference and sample arms, and recombined at the detector. The combination of reflected light from the sample arm and the reference light from the reference arm give rise to an interference pattern. OCT is different from conventional interferometry in that the interference occurs over a distance of micrometers (rather than meters).

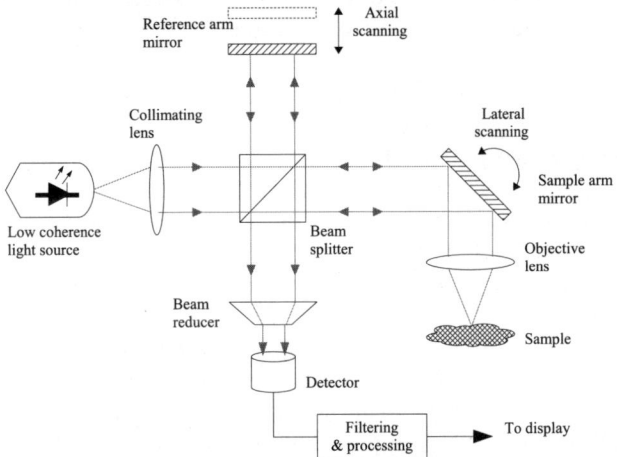

OCT detects the backscattered photons. By changing the length of the reference arm (i.e. scanning the mirror), light backscattered or reflected back from the sample combines with light reflected from the mirror to result in either a constructive or destructive interference [99]. If the length of the sample arm is also changed, then photons from different depths of the sample can also be detected.

Photon detection with OCT requires that the path-length difference of the photons coming from the sample arm and from the reference arm be smaller than the coherence length of the light source. The achievable depth resolution is governed by the coherence length of the light source, and the length is inversely proportional to the light's bandwidth. Over small distances the OCT signal is formed by single backscattered photons, but in deeper scattering samples the multiple-scattering effect (i.e. "echo" [100]) has to be considered. (Adapted from [95]).

"Occlusion spectroscopy" (or Occlusion Red Near-InfraRed Spectroscopy, O-RNIRS) (see [101, 102]) uses this concept of AC measurement, based on the

observation that a cessation of blood flow triggers a change of the optical characteristics of blood [103]. It was discover that when blood flow is suddenly stopped,
the average size of scattering particles in blood changes, allowing more light to
transmit through the tissue.

Before a stop of blood flow is imposed, red blood cell (RBC) aggregates less
due to the force of the ongoing blood flow. When a temporary over-systolic
pressure is introduced, there is a stop of blood flow, and the RBC starts to
aggregate. The increase in the size of the blood scattering particles cause a
change in the scattering coefficient, which is related to the mismatch between
the RBC and the plasma/ISF refraction index [102–104] (see Section 2.4.3).

Blood parameters such as hemoglobin, glucose, oxygen saturation, etc. influence the light transmission following over-systolic occlusion. For example, an
increased glucose concentration reduces the refractive index mismatch, while a
decreasing glucose concentration increases the refractive index mismatch, causing light transmission to decrease.

Hence, one way to measure the glucose value (using the above observation) is
by looking at the difference in light absorption of the tissue measured during the
systole and the diastole period. The systole and diastole are generally caused
by the pumping action of the heart (or heartbeat), but can be simulated by
tissue manipulation (i.e. applying artificial pressure to the tissue to cause a
volumetric change in the blood, e.g. by pressing on the finger). It was reported
that the glucose in arterialized whole blood correlated with glucose prediction
by O-RNIRS [103].

Yamakoshi et al [105] used the concept of AC measurement in "Pulse glucometry" technique, where Near-IR transmittance spectra was taken during normal
blood volume pulsations throughout the cardiac cycle. PLS calibration is used
to predict the BG values. Their approach differs to "occlusion spectrometry" in
that no pressure was applied to the tissue during measurement.

2.4.5 Raman Spectroscopy

Raman spectroscopy is a spectroscopic technique based on inelastic scattering
of monochromatic light [106]. There are generally two types of scattering[8]:

- elastic scattering – Also known as Rayleigh scatter, the incident photon
  "bounces" off the molecule with no gain or loss of energy, and thus the
  scattered photon has the same energy (frequency and wavelength) as the
  incident photons. In Rayleigh scattering, the dimension of the molecules or
  aggregates of molecules are significantly smaller than the wavelength of radiation [67].
- inelastic scattering – Also known as Raman scatter, the incident photon
  exchange energy with the molecule, leading to the emission of another
  photon with a different frequency than that of the incident photon [67,
  107].

---

[8] If a radiation is extinguished when it interacts with a molecule, then the process is
known as "absorption".

When the photon transfers energy to (or from) the molecule during an inelastic collision, the energy difference between the incident light (E_i) and the Raman scattered light (E_s) is equal to the energy involved in getting the molecule to vibrate. This energy difference is called the Raman shift [108]:

$$E_v = E_i - E_s$$

The loss (Strokes shift) or gain (anti-Stokes shift) of photon energy owing to the transitions of rotational and vibrational energy states within the scattering molecule causes the frequency of the scattered photon to shift up or down in comparison with original monochromatic frequency. The vibrational spectra produced as a result provide specific information about the chemical structure of the sample [65, 109], and reveals the interactions between the molecule and its local chemical environment [110].

Note that the observed Raman shifts are independent of the excitation frequency [67], and thus an excitation frequency may be chosen which is appropriate for a particular sample. However, the intensity of Raman scattered peaks generally falls off with decreasing frequency [109].

Most incident photons undergo elastic Rayleigh scattering, and about 0.001% of the incident light produces Raman signal [65]. Since this signal is very weak [108], special measures should be taken to distinguish it from the dominant Rayleigh scattering [106]. Using Raman spectroscopy for in vivo transcutaneous glucose measurement is difficult because whole blood and most tissue are highly absorptive and containing many fluorescent and Raman-active confounders [110].[9]

Instrumentation

Fig 2.8 shows an example of the Raman spectroscopy setup used in [111]. Typically, the instrument consists of:

1. Excitation source (laser)
2. Wavelength selector (bandpass filter, holographic gating)
3. Sample illumination assembly and light collection optics
4. Detector (photodiode array, CCD or photon-multiplier tube).

A sample is normally illuminated with a laser beam in the ultraviolet, visible or near infrared range. Scattered light is collected with a lens and is sent through the detector to obtain Raman spectrum of a sample.

To improve the Raman signal intensity, there are many other variations of Raman spectroscopy that have been developed, e.g. Surface Enhanced Raman Spectroscopy, Resonance Raman spectroscopy, to name a few (See [67]).

---

[9] Other site could be chosen, for example, the aqueous humor, which is relatively non-absorptive and contains few Raman-active molecules. Raman excitation wavelength in the Near-IR region can be used to minimise biological fluorescence and tissue damage [110].

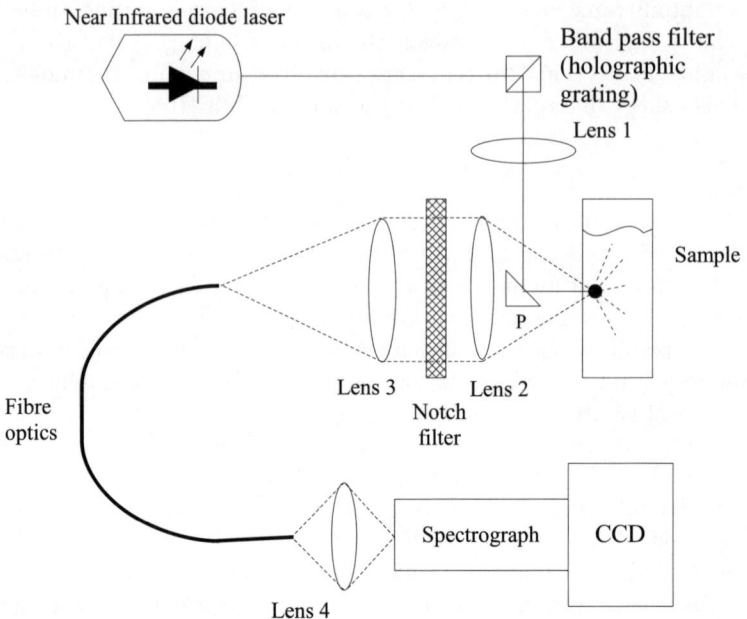

Fig. 2.8. Raman spectroscopy used in Berger et al's experiment (Adapted from [111])

2.4.6 Polarimetry

Polarimetric measurement of glucose concentration is based on the phenomenon of optical rotation (or optical activity), whereby a solution containing a chiral molecule rotates the plane of polarisation for linearly polarized light passing through it.

This rotation is the result of a difference in refractive indices η_L and η_R for left and right circularly polarised light traveling through the electron cloud of a molecule. It occurs because of the chirality or "handedness" of the molecule (i.e. the molecule has at least one centre about which its mirror image cannot be superimposed upon itself) [38].

In pure liquids, the angle of the rotation depends linearly on the optical pathlength L [dm], the density of the liquid D [g/ml] and a constant for the liquid $[\alpha]_\lambda^T$:

$$\phi = [\alpha]_\lambda^T \times L \times D$$

The constant $[\alpha]_\lambda^T$ is known as the specific rotation[10], for a wavelength λ [nm] and measurement temperature T [degree Celsius].

---

[10] Specific rotation $[\alpha]_\lambda^T$ is defined as the number of degrees of optical rotation ϕ observed when plane-polarized light is passed through a material with a path length of 1 decimetre and the sample concentration of 1 gram per 100 ml [65]. For solution, its unit is $[\,^\circ\ \mathrm{dm}^{-1}(\mathrm{g}/100\,\mathrm{ml})^{-1}]$. For pure liquid, the liquid's density replaces the solution concentration, i.e., the unit is $[\,^\circ\ \mathrm{dm}^{-1}(\mathrm{g}/\mathrm{ml})^{-1}]$.

Table 2.4. Optical activity

Optical Activity

A linearly polarised light can be regarded as an equal combination of right and left circularly polarised light with equal amplitude. As illustrated in [112],

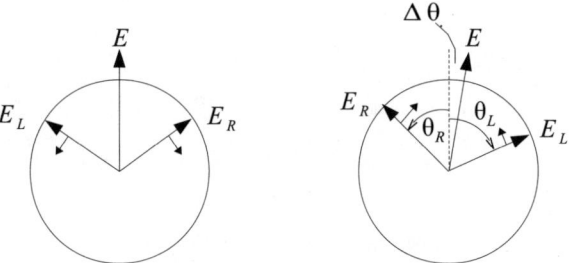

where $\theta_R = -\frac{2\pi L}{\lambda}\frac{c}{v_R}$, $\theta_L = \frac{2\pi L}{\lambda}\frac{c}{v_L}$, λ = wavelength of the light (in vaccum), c = speed of light, L = length of material, v_R and v_L = velocity of the right and left circularly polarised components in the medium. The angle of rotation is

$$\Delta\theta = \frac{1}{2}(\theta_L + \theta_R) = \frac{\pi L}{\lambda}\left(\frac{c}{v_L} - \frac{c}{v_R}\right)$$

In an optically active material, the two circular polarisations will experience different travelling velocities. This difference (also called optical activity strength) is characteristic of the material. Since the refractive index is defined as $\eta = c/v$, the angle of rotation of the light that passes through the material can be expressed as

$$\Delta\theta = \frac{\pi L}{\lambda}(\eta_L - \eta_R)$$

A circularly polarised light traces out a circle as the electric wave propagates [113]. The final polarisation is rotated to angle $\theta + \Delta\theta$. From this equation, the degree of rotation depends on the colour (frequency) of the light, the path length L and the properties of the material.

In a large aggregate of randomly oriented molecules, the net rotation by the material averages to zero. However, in the case where the molecule is chiral, the rotation is additive [109].

In solution, the equation is [114–116]:

$$\phi = [\alpha]_\lambda^T \frac{L \times C}{100}$$

where L = optical path length [dm] and C = concentration [g/100 ml] for a sample at wavelength λ [nm] and temperature T [degree Celsius]. If the wavelength of the light used is Yellow Sodium D line at 589 nm, and the temperature is

20°C, then the specific rotation can be written as $[\alpha]_{589}^{20} = +52.6°$ (see [65]) or $[\alpha]_{D}^{20} = +52.6°$ (for Sodium "D" line).

Glucose in the body is dextro-rotatory (rotates light in the right-handed direction) and has a specific rotation of $+52.6°$ dm^{-1}(g/ml)$^{-1}$ at the sodium D-line of 589 nm [38, 109].

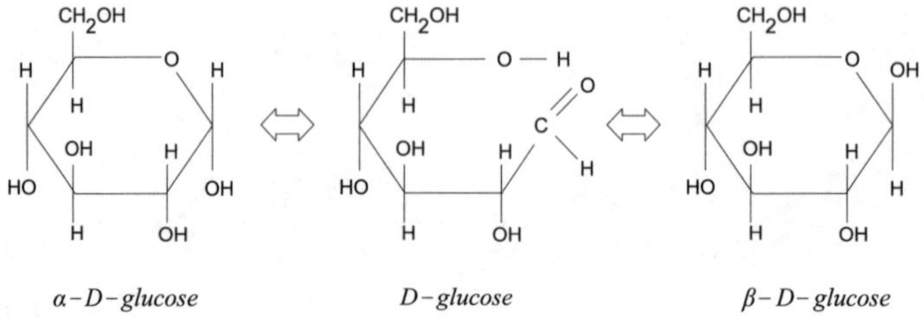

α−D−glucose   D−glucose   β−D−glucose

Fig. 2.9. Glucose is an equilibrium mixture of 2 enantiomers (i.e. nonsuperimposable mirror images of each other [65]): α-D-glucose and β-D-glucose. When pure α-D-glucose is dissolved in water, the specific rotation of the sample decreases from $+112°$ to $+52.6°$. When pure β-D-glucose is dissolved in water, the specific rotation increases from $+19°$ to $+52.6°$.

Generally, $[\alpha]_{\lambda}^{T}$ for glucose decreases with increasing wavelength across the visible light spectrum and has a rise in magnitude near the optical absorption bands of a particular molecule [38, 114]. The variation in the optical rotation of a substance with a change in the wavelength of light is called Optical rotatory dispersion (ORD) [65], and the observed rotation is

$$\phi = \frac{\pi}{\lambda} (\eta_L - \eta_R)$$

where λ = wavelength, η_L= refractive index of left-circularly polarised light, and η_R = refractive index of right-circularly polarised light.

Instrumentation

The general setup for polarimetric measurement is illustrated in Fig 2.10 and 2.11.

In the simplest form, the light travels from the source is linearly polarised by polariser 1. After passing through the sample, the polarisation plane will rotate due to the optical activity of the sample. The second polariser (also known as the analyser) is placed perpendicular to the first polariser. If there is no sample, then theoretically no light will be transmitter to the detector from the second polariser [117]. If an optically active sample is introduced, then the intensity of the light incident on the detector will be proportional to the square of the amplitude of the Electric-field passing through the second polariser. The E-field is proportional to the sine of the rotation angle φ by the sample [109].

An optical modulator (such as a Faraday rotator) can also be placed in between the sample and second polariser. The modulator causes the angle of the

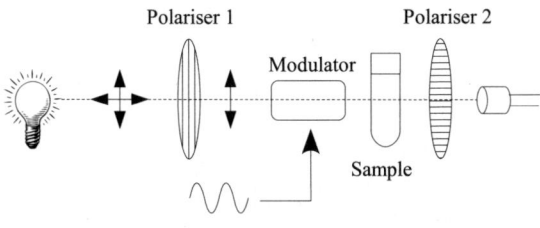

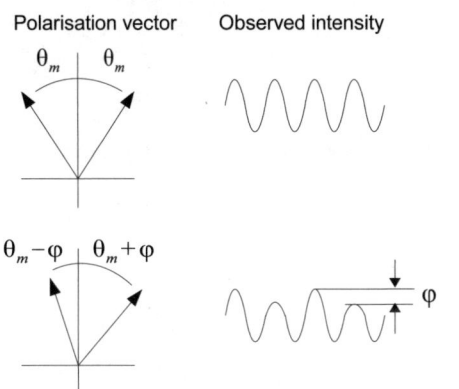

Fig. 2.10. Polarimetry glucose measurement (Adapted from McNichols et al [38]). Linearly polarised light is passed through a material, and the material "twisted" the polarisation angle of the light. This results in a varied light intensity when measured at the detector.

polarization vector to vary periodically [109]. The introduction of rotation of the linear polarisation vector of the light beam would eliminate amplitude variations due to fluctuations in the light source [118, 119].

Apart from measuring the light amplitude, the phase of the incident and transmitted light can be measured instead [118]. There are other implementations of the polarimetry system in existence, e.g. using a coil wound around the solution instead of a Faraday rotator to reduce cost [120], signal enhancement using digital closed-loop control of the polarimetric system to improve system reliability and stability [114], using optical heterodyne polarimeter [121, 122], or using the effect of Magnetic Optical Rotatory effect (MORE) where the optical rotations are multiplied on reflecting back through a medium in the presence of magnetic field [112]. MORE was used by Jang et al [123] to double the angle of rotation.

Obstacles to Polarimetric Glucose Measurement in Human

Human skin exhibit significant light scattering [124] and can depolarise the light passing through it. Hence it is not a feasible site for polarimetric measurement. The aqueous humour in the anterior chamber of the eye has relatively

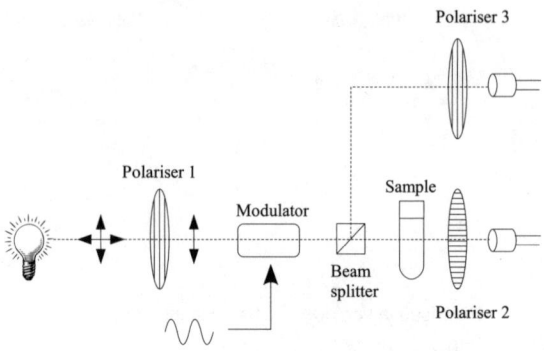

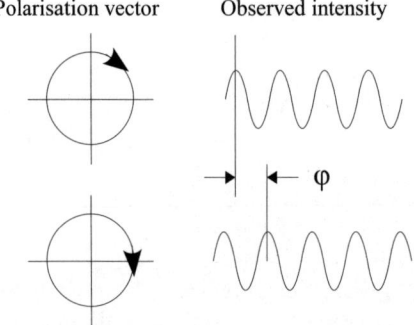

Fig. 2.11. Polarimetry glucose measurement by phase difference (Adapted from McNichols et al [38]). Instead of measuring the polarisation angle directly as light intensity, the angle is measured as a difference in the phase of incident and transmitted light.

low scattering properties, and has shown close resemblance of glucose levels in the blood [117], with a time lag of minutes [109, 114]. Nevertheless, glucose is not the only optically active component in the aqueous humour, and there are other potential problems with polarimetric glucose sensing in the eye including optical rotation due to the cornea (in addition to aqueous humor), birefringence of the corneas, and motion artifacts [38].

2.4.7 Photoacoustic

Photoacoustic (PA) refers to the generation of acoustic waves by modulated optical radiation [125]. The PA effect was discovered by A.G. Bell (1880) when he observed that audible sound is produced when chopped sunlight is incident on optically absorbing materials [75].

Most PA generation is due to "photothermal" (PT) effect, i.e. the heating of the sample after the absorption of optical energy [125]. PT heating of a sample is frequently produced with the use of laser beams.

In PA glucose measurement, the tissue is illuminated by pulses of low-energy laser set at a wavelength that is absorbed by glucose. The light pulses are focused to a relatively small region. When the light pulses are absorbed by glucose, the energy is converted into kinetic energy, which causes the temperature and pressure of the absorbing tissue to increase. This generates acoustic waves that radiate out from the region [126]. The PA waves can be detected using acoustic sensor such as microphone, piezoelectric transducer [127], or capacitance transducers. Other ways to detect the PA waves are, for example, shining a light (so-called a probe beam) onto the surface and detecting the deflection of the beam reflected from the surface, or interferometry, or detecting the refractive-index change associated with the PA pulse [128, 129].

The peak pressure of the PA wave can be described by:

$$P = K\frac{\beta v^n}{C_p}E_0\mu_a$$

where K = system constant, β = volumetric thermal expansion coefficient, v = sound velocity in the medium, n is between 1 and 2 (depending on experimental condition), C_p = specific heat capacity, E_0 = incident laser pulse energy, and μ_a = optical absorption coefficient [76].

When glucose concentration increases, C_p decreases while the acoustic velocity v increases. At glucose absorption wavelength, changes in PA signals are the result of changes in μ_a, v, and C_p [94]. Hence, the intensity of the PA waves can be related to the concentration of glucose in the absorbing tissue region.

The PA wave can be generated with:

- one excitation wavelength (e.g, Zhao & Myllylá [129] used wavelength of 905 nm and short low-energy pulse)
- over a range of excitation wavelength (e.g. Quan et al [130] scanned across the wavelength of 0.75–1.75 μm)

using

- point-by-point measurement – where the wavelength is shone on a spot on the tissue, and the PA signal is recorded.

  A range of wavelengths can also be scanned, with the corresponding PA signal for each wavelength recorded. Mackenzie et al [76] used 800–1200 nm wavelength at 7 ns pulse repeated at 10 Hz.

- multiplexing – the tissue is simultaneously excited by many wavelengths [85, 131].

  The multiplexed signal can then be transformed back into the conventional PA spectrum using suitable transform techniques (e.g. Fourier transform). (E.g. Spanner et al [132] used three wavelengths 834, 1304 and 1554 nm continuous-wave output powers at one focal point).

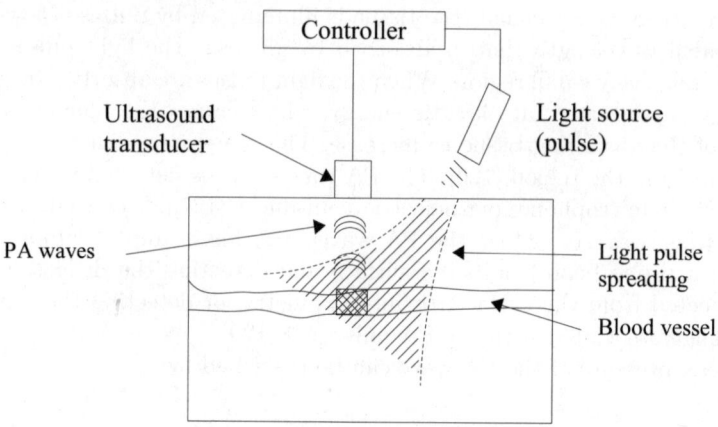

Fig. 2.12. Photoacoustic glucose measurement instrument setup, as used by Glucon Inc [126]

Fig 2.12 shows the instrumentation aspect of a photoacoustic glucose measurement system (Asprise^{TM} sensor) by Glucon Inc (Boston, MA, USA), described in [126] and [133].

PA response may still be interfered by other factors such as physiological change, blood circulation, water content and body temperature drift [128], but preliminary results from clinical testing showed promising result (See [131]). Readers interested in further details on photoacoustics are encouraged to read [125] and [128].

2.4.8 Light Diffraction

Measurement of glucose based on light diffraction has been reported in the literature. The technique used material that can swell or shrink when interacting with glucose [134, 135]. Light diffraction is governed by Bragg's law [134, 136]:

$$m\lambda = 2nD \cdot \sin\theta$$

where m = order of diffraction, λ = wavelength of the light in vaccum, n = average refractive index of the system (hydrogel, colloid), d = spacing between the diffracting planes/fringes, and θ = glancing angle between the incident light direction and the diffracting planes.

Glucose sensor constructed using this technique used:

- Crystalline colloidal array in hydrogel [134]

  Glucose oxidase and an enzyme were incorporated into a colloidal crystalline array (CCA). When glucose oxidase reacts with glucose, the enzyme became negatively charged, causing the hydrogel to expand. Because the brightly coloured CCA diffracts visible light according to the Bragg condition, any change in the size of the hydrogel can cause a change in the wavelength or colour of the diffracted light.

- Holographic sensor in hydrogel [135–137]

  This sensor is based on planar small-volume polymer hydrogels containing silver halide, which acts as a holographic recording material. The hydrogel contains the reflection hologram that reflects a narrow band of wavelengths when illuminated with white light. By incorporating ligands that interact with glucose (via competitive-binding), the hydrogel can change its swelling state upon analyte binding, which in turn causes a change in the wavelength of light diffracted by the holographic grating within the hydrogel. The wavelength is determined by the spacing between the fringes of the hologram (where an increase in the fringes increases would cause longer wavelengths to be diffracted). A holographic contact lens glucose sensor has been clinically trialled (see [137]).

2.4.9 Dielectric Spectroscopy

It was known that small changes in glucose concentration in blood can induce significant reactions in tissues involved in the metabolism of carbohydrate (such as liver, pancreas and red blood cell) [138]. One of the response of these tissues to increasing glucose concentration is a decrease of sodium and increase of potassium due to water movement from the tissue and red blood cell into the vascular system [138]. The variation of the electrolyte balance in the blood leads to a change in membrane potentials on the cellular level. Changes in membrane potentials can be recorded with dielectric spectroscopy.

In dielectric spectroscopy (or impedance spectroscopy), a small AC current is applied and the impedance of the tissue to the current flow is recorded as a function of frequency [139]. For non-invasive glucose monitoring, changes in skin impedance is measured. Skin impedance is sensitive to changes in membrane potential, which is influenced by the interaction of glucose with red blood cell.

Caduff et al [140] reported a non-invasive measurement of glucose using impedance spectroscopy. Fig 2.13 shows the equivalent circuit of the sensor mounted on the skin, attached to the impedance measuring network. The impedance of the sensor at a given resonance frequency depends on impedance changes within the human skin and underlying tissue, and also the skin temperature [140]. A frequency sweep across 1–200 MHz and temperature measurement were performed every minute to calculate the impedance minima $|Z|_{min}$. The sensor system was also incorporated into a wristwatch device.

Although the measured signal have a closer correlation to glucose changes in blood than to those in ISF, a few factors remain to be addressed, e.g. sweat and relocation of the sensor system (in order to obtain reliable glucose related signals), temperature variations, and spikes in registered signals due to movement [140].

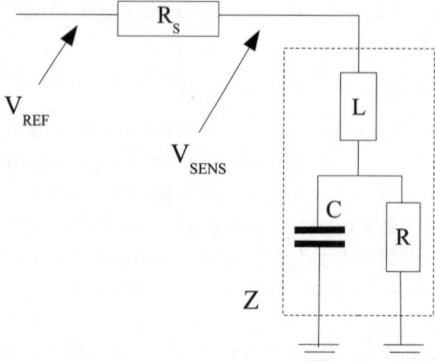

Fig. 2.13. Sensor for measurement of skin impedance (Adapted from Caduff et al [140]). R is the averaged resistance of the skin and underlying tissue. The voltage V_{REF} from a Voltage-Controlled-Oscillator is fed to a resistive-divider network consisting of a series resistor R_S and the sensor impedance Z. The measured sensor voltage is V_{SENS} and the impedance can be approximated by $Z_{\text{SENS}} \approx R_s \frac{V_{SENS}}{V_{REF} - V_{SENS}}$.

2.5 Amperometric Sensor Calibration

2.5.1 General Calibration

Since the current produced in the amperometric method is proportional to the level of glucose present, a calibration factor (or constant of proportionality) can be determined, and then used to convert all subsequent current measurements into an estimate of glucose concentration.

In implantable sensors, calibration methods based on "in-vitro", "in-vivo" and a combination of both have been suggested. In-vitro calibration determines the signal-to-concentration conversion factor using external glucose solutions with known concentration, while in-vivo calibration attempts to establish the relationship by matching the sensor signal to the glucose concentration in the blood. Examples of calibration methods are:

- "one-point in-vitro, one-point in-vivo" calibration [26];

  i.e. an in-vitro calibration is done before sensor implantation, followed by an in-vivo calibration after implantation. The in-vitro calibration is performed with the whole sensor system perfused in sterile phosphate buffered saline.

- "two-point in-vivo" [141];

  i.e. testing of the sensor's response using two in-vivo BGs and then applying a linear conversion to adjust the signal. Two points can be taken when an externally infused insulin or glucose caused BG to reach a new plateau or steady-state. The points are used to calculate an "in vivo sensitivity" coefficient which is expressed as nA/(mmol/l).

- "Wick" technique [142];

    i.e. the simultaneous measurement of glucose concentration in peripheral ve-
    nous plasma blood and in centrifuged wick fluid. The wick is made from a
    double-stranded cotton threads and is positioned subcutaneously to absorb
    the glucose, and then removed after an appropriate interval. The fluid ab-
    sorbed by the wick is then centrifuged, and the extracted glucose is correlated
    with those obtained from the plasma blood.

2.5.2 Regression Calibration

In contrast, MiniMed uses "Regression calibration" to calculate the calibration
factor, or rather, a set of calibration constants. Regression calibration is a modified form of linear regression. A regression slope is first calculated with an initial
fixed intercept of zero, by a linear regression of a minimum of four calibration
BG level entries, paired with corresponding valid sensor readings (called Isig
with unit in nA) for each 24-hour period (see [143]).

Mathematically, if we define:

x_i = i-th current measured by sensor (i.e. Isig) [nA]

$\bar{x}$ = background current of sensor in the absence of glucose [nA]

y_i = i-th calibration BG value (from glucometer) [mg/dl]

$\bar{y}$ = calibration BG value in the absence of sensor current [mg/dl]

Then least square regression line of y on x is

$$y = a + b_{xy} \cdot x \qquad \text{with} \qquad b_{xy} = \frac{S_{xy}}{S_x^2} \quad \text{and} \quad a = \bar{y} - b_{xy} \cdot \bar{x}$$

where

$$S_{xy} = \frac{\sum\limits_{i=1}^{n} (x_i - \bar{x})(y_i - \bar{y})}{n}$$

and

$$S_x^2 = \frac{\sum\limits_{i=1}^{n} (x_i - \bar{x})^2}{n}$$

with n-pairings of glucometer BG value and Isig each day, i.e. (x_1, y_1), (x_2, y_2),...,
(x_n, y_n) etc.

If the resulting slope b_{xy} is less than 7 mg/dl per nA, then an offset of 3 nA
is subtracted from each valid Isig, and the regression slope is re-calculated. The
BG level is then estimated from each Isig by:

$$\text{BG}_{estimate} = \begin{cases} b_{xy} \cdot \text{Isig}, & \text{if } b_{xy} \geq 7\,\text{mg/dl per nA} \\ b_{xy} \cdot (\text{Isig} - 3), & \text{if } b_{xy} < 7\,\text{mg/dl per nA} \end{cases}$$

The calibration BG level entries are based on BG measurements taken using
a conventional glucometer. The user enters the glucometer readings into CGMS

every day, while the monitor is in use. Isig values are deemed to be valid if they are within the range 10–100 nA. This calibration method can be viewed as a multi-point in-vivo calibration, with the exception that MiniMed has presented it for use in an "offline" manner, rather than for real-time estimation.

2.5.3 GlucoWatch Calibration

GlucoWatch is a wrist-watch device containing the electronic circuits and amperometric biosensors (called AutoSensor). The working electrode (WRK) is made of screen printed layer of Pt/C composite, while the reference (REF) and counter (CNT) electrodes are made of Ag and Ag/AgCl screen-print layers. The AutoSensor contains two hydrogel discs, which serve both as the electrolyte and reservoirs into which the glucose is collected [144]. Glucose oxidase is dissolved in these hydrogel discs. The glucose sample are first extracted from the skin via reverse iontophoresis into the hydrogel discs, and then measured by amperometric biosensor. The data are then processed prior to display.

The amount of glucose extracted during a 3-min iontophoresis cycle is estimated to be 50–500 picomol, which corresponds to 2.5–25 μM of glucose concentration in each hydrogel pad [144].

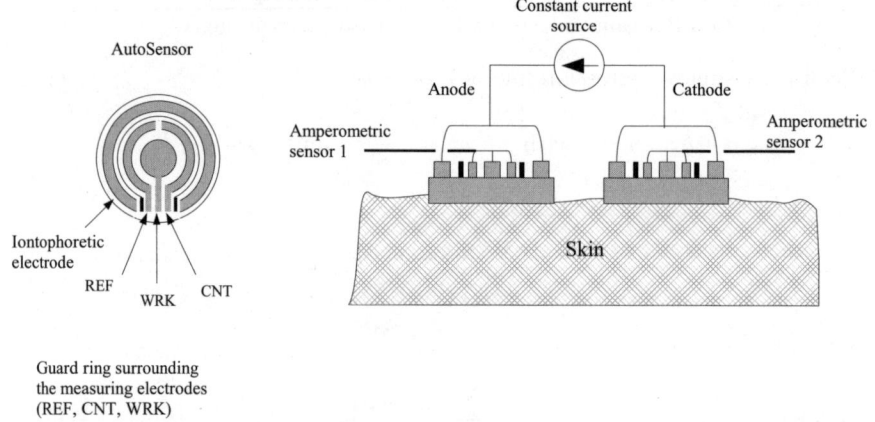

Fig. 2.14. A simplified view of the GlucoWatch mechanism (Adapted from [145] and GlucoWatch Patent US6233471 B1 [146])

The sequence of events in the operation of GlucoWatch are (as explained in [144]):

1. Iontophoresis current of 0.3 mA is delivered for 3 min to collect glucose at the cathode. The concentration of H_2O_2 increases in the hydrogel during this period.
2. Biosensors are biased at 0.42 V against the Ag/AgCl electrode. This low potential of 0.42 V for the detection of H_2O_2 improves selectivity towards glucose.

3. The current at the cathode is integrated for 7 min during which the concentration of H_2O_2 in the gel is depleted.
4. The polarity of the iontophoresis current is then reversed, and the whole process is repeated.

The integrated currents from the two biosensors are summed and input to a signal processing algorithm, where the GlucoWatch Biographer filters the current signals and then estimates BG from the signal. As described in [146], the signal processing consists of:

- Baseline background signal processing

  "Baseline background" is the current (nA) generated by the sensor independent of the presence or absence of the analyte of interest. This baseline background can vary with time, temperature and other variable factors. To remove or correct background information present in the sensor raw signal, a subtraction method can be used:

  $$i(\tau) = i_{raw}(\tau) - i_{bckgd}(\tau)$$

  where $i_{raw}(\tau)$ = raw signals measured by sensor [nA] at time τ, $i_{bckgd}(\tau)$ = background current [nA], $i(\tau)$ = corrected current [nA], and τ = time after activation of the sensor. To compensate for temperature, a different background current can be used:

  $$i_{bckgd} = A \exp\left(\frac{-K_1}{T}\right)$$

  where A = a constant, K_1 = "Arrhenius slope" – a measure of how sensitive the current is to changes in temperature, T = temperature in °Kelvin. [11] The temperature corrected background current is thus:

  $$i_{bckgd,Correct} = i_{bckgd,t=\tau_0} \exp\left[-K_1\left(\frac{1}{T_\tau} - \frac{1}{T_{\tau_0}}\right)\right]$$

  where $i_{bckgd,Correct}$ = temperature corrected baseline current, $i_{bckgd,t=\tau_0}$ = baseline current at $t - \tau_0$, T_i = temperature at $t = \tau$

- Sensor calibration

  The calibration can be carried out by giving a reference BG measurement (BG_{cal}, in [mg/dl]) to work out the conversion factor b_{gain} in [mg/dl per nC]:

  $$b_{gain} = \frac{BG_{cal} + \rho}{E_{cal} + \delta}$$

  where E_{cal} = blank-subtracted smoothed sensor signal (nC) at calibration, ρ = calibration offset [mg/dl], δ = offset calibration factor constant [nC] which

---

[11] K_1 can be found out by plotting the natural log of i_{bckgd} versus $\frac{1}{T}$. The slope of the function is $-K_1$.

can be calculated using standard regression analysis. Blank subtraction is another technique described in [146] where the signal from the inactive sensor reservoir (i.e. blank, no glucose loading) is used to compensate for interference from other analytes in the active sensor reservoir. The offset factor δ is used to account for a non-zero signal at an estimated zero BG. This is needed because sometimes a non-zero intercept is obtained in the correlation between signal and reference glucose value.

To compensate for time-dependent decline in sensor signal, additive decay parameters α_i and multiplicative decay parameters ϵ_i can be included:

$$b_{gain} = \frac{\mathrm{BG}_{cal} + \rho - \alpha_i \cdot t_{cal}}{(1 + \epsilon_i \cdot t_{cal})\, \mathrm{E}_{cal} + \delta}$$

where t_{cal} is the time when calibration is performed, α_i and ϵ_i are time-dependent signal decline and can have multiple time segments (i.e. $i = 1, 2,$ or 3). The BG at time t can be estimated using

$$\mathrm{BG}_{est} = b_{gain} \left[(1 + \epsilon_i \cdot t)\, \mathrm{E}_t + \delta \right] - \rho + \alpha_i \cdot t$$

• Mixture of Expert (MOE)

The MOE algorithm has been reported in [147] and proposed for GlucoWatch biographer. MOE is described as a generalized predictive data analysis method that superimposes multiple linear regressions, along with a switching algorithm to predict outcomes. For GlucoWatch, the MOE inputs are the elapsed time (time), integrated current (I_{avg}), blood glucose values at the calibration point (BG_{cal}) and a calibrated sensor signal (I_{cal}). The BG is estimated using the equations

$$\mathrm{BG} = \sum_{i=1}^{3} w_i \cdot \mathrm{BG}_i$$

where w_i are weighting factors and BG_i are individual experts. Each expert is described as a linear summation of appropriate input parameters P_j, plus a constant (see [147]):

$$\begin{aligned} \mathrm{BG}_i &= \sum_{j=1}^{4} a_{ij} P_j + z_i \\ &= a_{i1} \cdot \text{time} + a_{i2} \cdot \mathrm{I}_{avg} + a_{i3} \cdot \mathrm{I}_{cal} \\ &\quad + a_{i4} \cdot \mathrm{BG}_{cal} + z_i, \qquad i = 1, 2, 3 \end{aligned}$$

The weights w_i are determined from

$$w_i = \frac{\exp(\phi_1)}{\exp(\phi_1) + \exp(\phi_2) + \exp(\phi_3)}, \qquad i = 1, 2, 3$$

where

$$\phi_i = \sum_{i=1}^{4} \alpha_{ij} P_j + \zeta_i$$

$$= \alpha_{i1}\dot{\text{time}} + \alpha_{i2} \cdot I_{\text{avg}} + \alpha_{i3} \cdot I_{\text{cal}}$$
$$+ \alpha_{i4} \cdot BG_{\text{cal}} + \zeta_i, \qquad i = 1, 2, 3$$

To determine the unknown coefficients (a_{ij}, z_i and α_{ij}, ζ_i), a technique called Expectation Maximization is used as a two-step process: (1) The weights associated with each data point are fixed, and the parameter values are optimised; (2) the parameter values are fixed and the weights are optimised. These two steps are iterated until convergence is reached [147].

For a good technical review of GlucoWatch Biographer, see [56].

2.5.4 Differences Between the MiniMed's CGMS^{tm} and Cygnus GlucoWatch Biographer®

Both CGMS and GlucoWatch use the same principle of current generation from the oxidation of hydrogen peroxide. The hydrogen peroxide is created by the enzymatic action of glucose oxidase on glucose. However, the signal processing in GlucoWatch requires a more comprehensive mathematical formulation (as seen above) than the simpler Finite Impulse Response filter used in CGMS (as reported in [148]). This is due to the differences in physiological pathways, and the concentration of the glucose being measured by the two devices. Nevertheless, both products require users to enter a glucometer reference value to perform a single or multi-point calibration.

2.6 Clarke's Error Grid Analysis (EGA)

Self-monitoring of blood glucose (SMBG) has become important in the management of diabetes mellitus. In order to assess the accuracy of a particular glucometer, Clarke et al devised an analysis method that evaluates the accuracy of a particular system over the entire range of BG value, and the clinical and statistical significance of the system's accuracy. It was designed to take into account the absolute value of the glucometer's glucose measurement, the absolute value of the reference blood glucose value, the relative difference between these two values, and the clinical significance of this difference [149].

EGA defines the x-axis as the reference blood glucose and the y-axis as the value generated by the monitoring system. The diagonal represents the perfect agreement between the two. Data points situated above or below the diagonal represents overestimates and underestimates, respectively.

EGA is based on the assumptions that (from Clarke et al):

1. the target blood glucose range, or the range of glucose values that patients were taught to attain and maintain is 70–180 mg/dl (3.88–10 mmol/L);
2. patients will attempt to correct BG readings that are above or below the target range, but not those readings within the target range;

3. corrective treatment by the patient is considered inappropriate if such treatment causes the BG reading to fall outside of the target range;
4. failure to treat BG values <70 or >240 mg/dl is inappropriate.

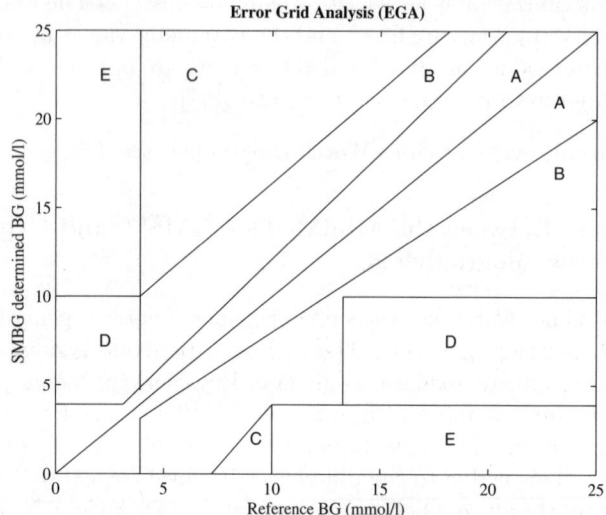

Fig. 2.15. Error Grid Analysis (Clarke et al)

EGA is divided into five zones of glucose estimations (Fig 2.15):

- Zone A – Glucose values deviate from the reference by no more than 20%, or glucose values are in the hypoglycemic range (<70 mg/dl or 3.88 mmol/l) when the reference is also <70 mg/dl. Values falling within zone A are clinically accurate in that they would lead to clinically correct treatment decisions;
- Zone B – Values that deviate from the reference by >20% but would lead to benign or no treatment (based on the assumptions described above);
- Zone C – Values that would result in overcorrecting acceptable BG levels, and such treatment might cause the actual BG to fall outside of 70–180 mg/dl (3.88–10 mmol/l) range;
- Zone D – Actual BG values are outside of the target range, but glucometer values are within the target range. This would cause "Dangerous failure to detect and treat" errors;
- Zone E – Reference BG values and glucometer BG values are opposite to each other, and corresponding treatment decisions are therefore opposite to that called for. This is the "erroneous treatment" zone.

In summary, values in Zone A and B are clinically acceptable, while values in zones C, D and E are potentially dangerous and are clinically significant errors.

EGA has since been used quite extensively to compare the accuracy of glucometers (commonly against a reference meter such as Yellow Springs Instrument (YSI) whole blood glucose analyser).

2.7 Insulin Infusion

2.7.1 Fate of Insulin

Physiological insulin delivery from the endocrine pancreas is a continuously regulated process, which responds to changes in substrate concentration in a matter of seconds [150]. In a normal individual, endogenously produced insulin is secreted directly into the portal venous circulation, and undergoes partial (40% to 80%) extraction by the liver. After this clearance, it is diluted into the systemic insulin pool [151,152], and is distributed within the plasma as "free" (unbound) insulin. This biologically active free insulin is then distributed by diffusion into the extravascular compartment to reach insulin receptors on the surface of the target cells. The interaction of insulin with its receptor enables fuel (mostly carbohydrate and amino acid) to enter cells and activate enzymes for the storage or metabolism of those fuels. Insulin that does not combine with receptor cells is degraded mainly in the liver. Insulin is also cleared from the plasma by the kidney.

2.7.2 Insulin Type Factor

In contrast to endogenous insulin, some exogenously infused insulins require further intermediate steps before being active at the target cells. Because insulin is more readily absorbed as dimers and monomers, insulin preparations that contain the larger hexamers (such as regular injectable insulin) must first dissociate into dimers and monomers to be absorbed [28].

Conversely, insulin preparations that contain monomers and dimers are absorbed more rapidly [28]. Shorter-acting insulin neutral and rapid-acting insulin Lispro are some examples of readily absorbable insulins. Figure 2.16 shows time of activity of some commonly used insulin.

Exogenously infused insulin is degraded at the usual sites: liver, muscle, and kidney. However, the kidney plays a larger role in clearing exogenously administered insulin (than endogenous insulin), and it clears approximately 30–80% of exogenous insulin from the systemic circulation.

2.7.3 Routes of Insulin Administration

As insulin is normally cleared rapidly from the plasma, circulating insulin levels and its ultimate delivery to the target tissues depends mainly on the entry of insulin to the circulation.

In a healthy individual, insulin enters the circulation mainly from pancreatic release. But if there is an impairment of endogenous production (i.e. diabetes), exogenous insulin must be administered. Exogenous insulin can be administered by different routes:

- Intraperitoneal (i.p.).
- Subcutaneous (s.c.).

Time of Activity of Human Insulins*

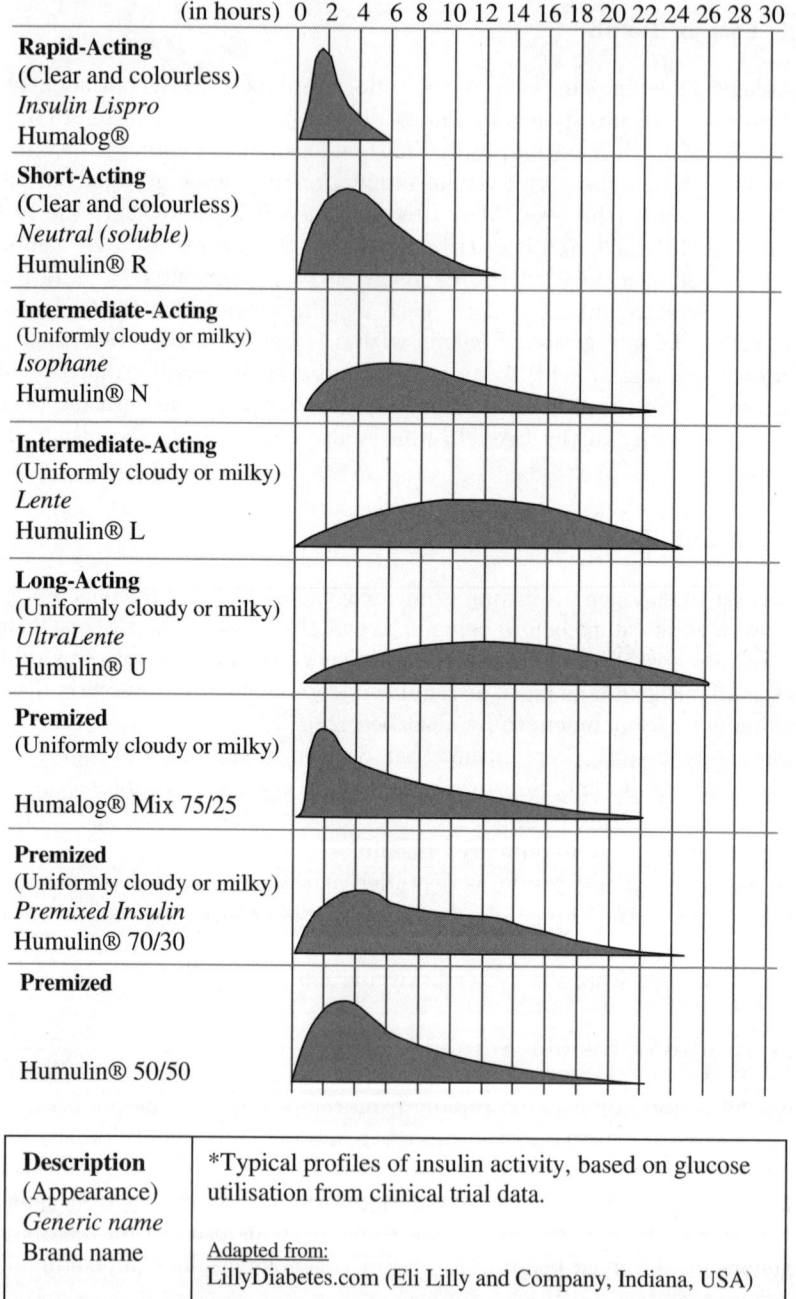

Fig. 2.16. Human insulin action activity chart. (Source: Eli Lilly & Co., Indiana, USA).

- Intravenous (i.v.).
- Intramuscular (i.m.).
- Alternative non-invasive routes, such as oral, nasal, pulmonary or transdermal/transcutaneous.

The intraperitoneal, subcutaneous and intravenous routes are the three most common routes for continuous administration of insulin.

Intra-peritoneal Delivery

In a normal individual, insulin is first delivered to the hepatic circulation by the pancreas. Hence, it follows that delivery of insulin to the liver should produce more normal metabolism of glucose and other metabolites, than peripherally delivered insulin.

The peritoneum is a possible site for insulin administration [150] because the visceral peritoneal membrane has the same blood supply as the small and large bowel, and is drained via the portal vein to the liver. Therefore, the delivery of insulin intraperitoneally should, theoretically, have the advantage of insulin being absorbed preferentially into the portal system, producing a therapeutic state more closely resembling the normal physiology.

Numerous studies on insulin-dependent diabetes mellitus patients have demonstrated that insulin delivered intraperitoneally is rapidly and predictably absorbed, with most of it going into the portal system [153]. Intraperitoneal delivery:

- provides sufficient insulin to the liver to control hepatic metabolism;
- reduces peripheral insulin levels, and the risks of arteriosclerosis, and other adverse effects of hyperinsulinemia (e.g. increased risk of developing Alzheimer disease [154, 155]);
- results in carbohydrate and particularly lipid metabolism that more closely mimics the normal physiological state, than those produced by s.c. injections of insulin.

Nevertheless, intraperitoneal insulin delivery requires an insulin pump to be surgically implanted in the peritoneum. Any mechanical failure to the pump poses risks to the patient by the need to go through surgery for pump removal.

Subcutaneous Delivery

S.c. injection is the most preferred route by patients to receive their daily insulin doses, either by direct injection at the site, or via insulin pump therapy.

When insulin is injected subcutaneously, it forms a depot at the injection site. The compound then transforms to an absorbable state through dissociation and dissolution [156], and diffuses from the depot into the circulation. The diffusion process is slow, taking 50–90 min in general [28]. Similar to endogenous insulin, the injected insulin is distributed within the plasma as "free" (unbound) insulin, diffusing further into the extravascular compartment to reach the target cells.

Some shortcomings of s.c. delivery are:

1. The s.c. absorption rate is highly unpredictable, mainly attributed to changes in blood flow around the injection site [151]. In most insulin-treated subjects, part of the circulating insulin is reversibly bound to insulin antibodies, which changes the free insulin levels unpredictably;

2. Variation exists in insulin absorption due to factors such as depth and location of injection, exercise, and vascularity of the tissue. These are common causes of unpredictable wide swings in blood glucose levels;

3. Although the insulin peaks can be adjusted to match food absorption by giving subcutaneous insulin in advance of meals, the difficulty of estimating when exactly to eat increases when insulin absorption could vary [153];

4. As s.c. insulin forms a depot at the injection site, once injected, any excess insulin cannot be withdrawn from the injection site. This may lead to hypoglycaemia if the patient does not ingest carbohydrate after the injection;

5. An inflammatory reaction might develop at the injection sites, necessitating frequent rotation of injection sites;

6. There is a long lag time between the injection of insulin and its appearances in the plasma (30 min to 1 hour).

Intravenous Delivery

Unlike subcutaneously injected insulin, intravenously administered insulin, especially as the soluble form, enters the bloodstream immediately, and is rapidly absorbed, thereby acting faster. Since i.v. insulin is usually injected into the veins at a peripheral site, and not directly into the hepatic circulation, the insulin avoids the first pass through the liver, and this effectively leads to an increase of 50% of insulin concentration, compared to i.p. or endogenous insulin availability.

Peripheral hyperinsulinemia may be a consequence of this route of infusion [16], and is not desirable, as emphasized by the mounting evidence that circulating insulin levels may be directly related to the risk for atherosclerosis [153].

The primary shortcoming of i.v. insulin administration is the requirement for venous access, which usually requires medical supervision.

Continuous Versus Intermittent Intravenous Delivery

Intravenous insulin may be given as either repeated boluses or continuous infusion. The bolus ensures insulin absorption, and permits rapid adjustment of insulin to suit the patients needs. However, the effect of each bolus of short-acting insulin dissipates within 30 minutes to 1 hour, necessitating frequent BG measurement and the delivery of insulin bolus. Also, it is possible for the patient to form anti-insulin antibodies, and thus be immunized against the intermittently delivered insulin.

In contrast, continuous intravenous insulin delivery allows a longer period between BG measurement, and permits fine-tuning of BG levels to within a chosen range, although it may increase the risk of hypoglycaemia.

Other Delivery Routes

The latest in insulin delivery technique has focused on non-invasive delivery:

- Oral – Capsules containing insulin for oral administration (or so-called Emisphere[tm] delivery agent combination for example, marketed by Emisphere Technologies Inc. (Tarrytown, N.Y.)).

- Nasal – The inhaled insulin delivery system provides insulin as a dry powder inhaled through the mouth directly into the lungs from where it passes into the bloodstream.

- Transdermal - Delivery of insulin through a transdermal "patch".

   One product is called "U-Strip" by the founding company Encapsulation Systems Inc (http://www.encsys.com). The device uses ultrasound to enhance transport of insulin through the skin pores (See US Patent 6,908,448). It is wearable and portable, and it features programmable drug delivery through its drug sonic applicator device. Another rivaling company is Sontra Medical Corporation (http://www.sontra.com) (See [58,60]).

   The advantages with insulin delivered through transdermal route is that the insulin suffers no degradation (in contrast to orally administered insulin that can be destroyed by the GI tract) and no painful needle stick is required.

Other Factors/Mode of Delivery

Apart from the type and infusion route of insulin, insulin action is also influenced by factors such as:
1. insulin resistance
2. insulin sensitivity — the sensitivity of the insulin receptors
3. the action of the counter-regulatory hormones,
4. physical exercise and diet
There has also been a report that oscillatory delivery of exogenous insulin with a period of 120 minute into the systemic circulation leads to a greater reduction in the plasma glucose concentration as compared to constant delivery [157]. This phenomenon was discovered in normal weight, non-diabetic subjects. Its implication in diabetic subjects is yet to be determined.

2.8 Conclusion

Various BG measurement methods exist, each with their advantages and disadvantages. Although the choice of a method depends on the target population and the clinical setting, it is undoubtedly that non-invasive method is the most preferable choice, and the way forward for the future. But until non-invasive glucose sensing techniques have reached maturity, minimally-invasive methods would remain the choice for longer-term continuous glucose monitoring. Invasive glucose monitoring remains in use only in clinical research settings (and not for day-to-day use by the general public).

Similarly, for the insulin infusion, the choice of insulin type and administration routes depends on the target population. The subcutaneous route is the preferred route for administration in ambulatory patients, despite possible discomfort, while invasive route is used mainly in clinical routine (due to its faster action). There are progress being made in non-invasive insulin delivery, and commercial products have started to appear. With the emergence of non-invasive insulin delivery techniques, the future of pain-free delivery appears hopeful.

2.9 Summary

This chapter discussed the three categories of BG measurement techniques, i.e. invasive, minimally-invasive and non-invasive techniques. The methods in each category were described, as well as their advantages and disadvantages. The chapter also paid particular attention to one glucose sensing method – amperometry – from how the signal is obtained from the sensor, to the steps involved in converting the signal to BG level estimations. Amperometric glucose sensor has reached successful commercialisation ahead of other measuring techniques, and this progress would certainly help in advancing glucose control by providing the necessary tools for further advancement in the field. This chapter also presented the types of insulin that were commonly used, and the effects of the routes of delivery on the insulin activity. Non-invasive delivery of insulin remains the ultimate goal by many to achieve totally pain-free insulin delivery.

3

Glucose Control: Patient Dynamics

3.1 Introduction

The role of the control algorithm in a closed-loop insulin delivery system is to regulate the patient's BG level, replacing the intrinsic glucose regulatory function, which is abnormal in diabetics. To develop an effective algorithm, a knowledge of how glucose is intrinsically regulated in a healthy person is essential.

The human pancreas has between 1 and 2 million islets of Langerhans. These islets contain three major cell types: alpha, beta and delta. The beta cells constitute about 60% of the cells, and secrete insulin. The alpha cells, about 25% of the total, secrete glucagon. The delta cells, about 10% of the total, secrete somatostatin. The remaining 5% of cells are made up of other cell types which secrete hormones of uncertain function [4]. Insulin and glucagon play the most important roles in the glucose-regulatory system.

The glucose-regulatory mechanism is not an isolated system, but has connections with many other metabolic pathways in the body.

This chapter reviews the various mechanisms involved in the glucose-regulation in normal individuals, diabetic patients and critically ill patients.

3.2 Intrinsic Blood Glucose Regulation

In a normal healthy person, blood glucose concentration is controlled to within a narrow band, usually between 3.5 mmol/l and 5.6 mmol/l in the fasting state. During fasting, insulin secretion is reduced to a basal level, and glucagon is released to allow the liver to:

- mobilize glucose from its glycogen stores (glycogenolysis) and
- synthesize glucose from amino acid (gluconeogenesis).

In addition, when insulin levels are low, the uptake of glucose by muscle is minimized, and there is lipolysis of stored fats to release free fatty acids. When fasting persists longer than 12 to 18 hours, these free fatty acids then become the main energy substrate, used by essentially all tissues of the body, except

F. Chee & T. Fernando: Closed-Loop Control of Blood Glucose, LNCIS 368, pp. 49–57, 2007.
springerlink.com

the brain. Gluconeogenesis still supplies glucose for obligatory glycolytic tissues, notably the brain.

This mechanism effects a stable fasting blood glucose concentration so that the brain, which has no energy stores, has a sufficient supply of nutrients for normal activity. Glucose is an essential nutrient for the brain, retina, and germinal epithelium of the gonads. Insulin is always present, and a low level of circulating insulin regulates the rate of lipolysis, glucose transport and gluconeogenesis at all times [158].

When a person prepares to eat a meal, two phases of insulin secretion occur: an anticipatory phase (first phase) and a glucose-sensitive phase (second phase).

In the anticipatory phase, the sight of food and the first bite of a meal cause the brain to send signals to the pancreas. These signals cause the pancreas to release insulin into the hepatic circulation (Fig 3.1). Once the insulin is in the hepatic circulation, the liver stops breaking down glycogen into glucose.

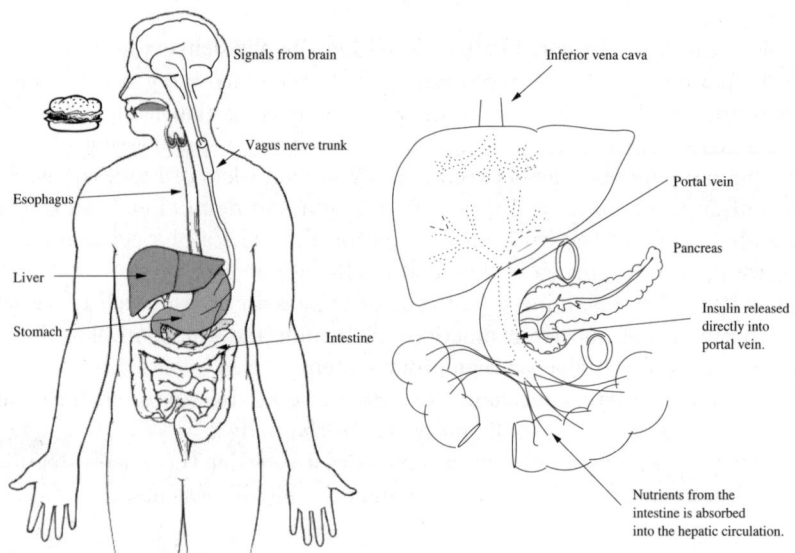

Fig. 3.1. Anatomy of the intrinsic blood glucose regulation mechanism

As the food enters the stomach, the release of insulin is further facilitated by gastrointestinal hormones. These hormones increase the sensitivity of the islet cells to glucose [159]. As nutrients are absorbed into the circulation, the glucose-sensitive phase begins, and there is continuous secretion of insulin. These two phases are sometimes termed the biphasic response of insulin secretion.

After absorption of all the carbohydrates, the feedback system for control of blood glucose returns the glucose concentration rapidly back to the control level, usually within 2 hours.

Although glucose is an important physiological stimulant of insulin secretion, nutrients other than glucose, particularly amino acids, are also capable of stimulating insulin release. Nevertheless, amino acids stimulate a much greater insulin response when accompanied by hyperglycaemia, than the modest degree expressed in the presence of normal plasma glucose level.

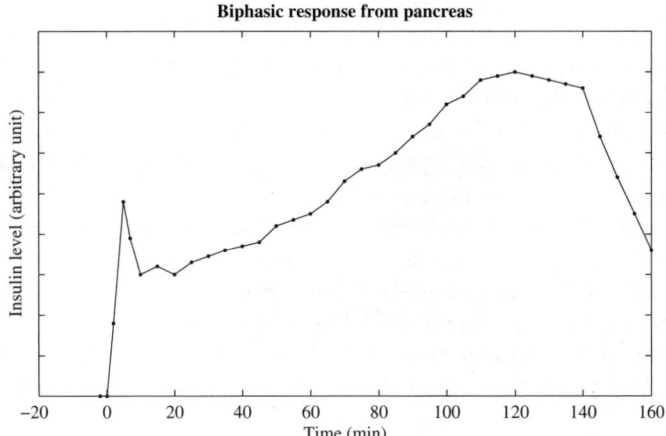

Fig. 3.2. Biphasic insulin response (illustration)

The biphasic response (Fig 3.2) is also observed in the plasma compartment during glucose clamp studies, where a square wave of hyperglycaemia is introduced intravenously under normal conditions in a normal individual. The first phase of the insulin response consists of a rapid rise in the insulin level during the first 10 min, with a peak response at 4 min; the plasma insulin level then falls and reaches a nadir at 10 min [160]. After 10 min, the plasma insulin level is then observed to increase gradually according to the degree of hyperglycaemia and persists for the duration of the stimulus [161].

3.3 Diabetic Patients

In patients with Type I and Type II diabetes mellitus, beta cells have either been partially (as in Type II) or completely (as in Type I) destroyed. The destruction of beta cells results in reduced or no insulin secretion, and also the disappearance of the first phase (anticipatory phase).

With no anticipatory phase, the liver does not receive the message to stop breaking down glycogen into glucose, resulting in continued hepatic glucose production. Added the glucose absorbed from the meal, and a lack of insulin release, hyperglycaemia ensued [162]. The significance of the first phase release has been demonstrated in [163] where a 15-minute delay in initiation of insulin delivery, relative to the start of the meal resulted in a greater meal-induced hyperglycaemia in patients.

The major difference between Type I and Type II diabetes mellitus is that Type I diabetic patients cannot survive without exogenous insulin, contrary to

Type II patients who suffer from either a reduced (but present) insulin secretion, or abnormal insulin response (increased peripheral insulin resistance) or both. Hence, Type I diabetes is also known as insulin-dependent diabetes mellitus (IDDM), where as Type II is also known as non-insulin-dependent diabetes mellitus.

The first phase insulin release is somewhat harder to achieve by a closed-loop insulin delivery system in an automatic manner. This is because the anticipatory phase in a normal individual is initiated by neuro- and gastro-intestinal signals, which are not measurable by a glucose sensor. The word "automatic" is used, because in open-loop delivery systems, this signal can be given by the patient by injecting insulin boluses manually before meal.

A solution to "artificially" achieve the first phase insulin release using a closed-loop system would be to use sensors that have a very fast response time (e.g. 1 min at least), and to immediately deliver insulin when BG level starts rising (with the assumption that the rise of BG level signifies the start of a meal (as inferred from [163])). In fact this method has been used in the early external artificial endocrine pancreas clinical experiments, such as those reported in [17] and [16].

As for the second phase insulin release, the matching of the insulin dose to the blood sugar intake depends on:

- A knowledge of how much glucose is ingested. In a healthy pancreas, beta cells also function as "fuel-sensors" and are capable of adapting the rate of insulin secretion to variations in plasma glucose level.
- The responsiveness of the insulin receptors on target cells to enable glucose to enter and be utilised by the cells.

3.4 Importance of BG Control

It is important for BG level to be kept within the normal range, because:

1. a high glucose concentration exerts an osmotic pressure in the extracellular fluid, and causes cellular dehydration. This excessive BG level causes loss of glucose through urination (glycosuria), leading to osmotic diuresis that depletes the body further of fluids and electrolytes.

2. too low a BG level carries the risk of hypoglycaemic coma. The BG level should not drop below a certain level because glucose is the only nutrient that can be used for energy by the brain, retina, and germinal epithelium of the gonads.

3. too high a glucose concentration (>11.1 mmol/l) can affect wound healing and interfere with human neutrophil function [8].

4. therapy that maintains BG level at below 11.9 mmol/l improves the long-term outcome in diabetic patients with acute myocardial infarction [9].

5. pre-prandial blood glucose concentrations between 3.9 mmol/l and 6.7 mmol/l, and postprandial concentration of less than 10 mmol/l was found to delay the onset of diabetic microvascular complications [11]. These microvascular complications included retinopathy (visual impairment), vision-threatening lesions, nephropathy (kidney disease) and neuropathy (nerve damage).

Complications not only occurred in diabetic patients, but also has an impact on critically ill patient population:

1. A recent finding (see [9]) has shown that the use of intensive insulin therapy to maintain BG at a level that did not exceed 110 mg/dl (6.1 mmol/l) substantially reduced mortality and morbidity in critically-ill patients in the ICU (8.0% with conventional treatment to 4.6% with intensive insulin therapy).
2. The finding also reported that a pronounced hyperglycaemia in critically-ill patients, even those who have not previously had diabetes, may lead to complications in such patients [9].
3. Patients with mean glucose concentrations >11.1 mmol/l within 36 h following surgery were more likely to develop infectious complications than their counterparts who were under better glycaemic control [14].

All these findings demonstrated that apart from the administration of exogenous insulin being essential to control the blood glucose concentration and to maintain homeostasis in patients, the tightness of such a control is crucial in avoiding long-term complications (in diabetic patients) and infectious complications (in critically ill patients).

3.4.1 New Findings

According to [164], it is believed that sensory nerves of pancreas release a neuropeptide called "substance P" which is responsible for ensuring that islet cells produce the right amount of insulin. On lacking of substance P, islets cells (in mice) overproduce insulin, leading to insulin resistance and ultimately islet-cell death. By injecting substance P to diabetic mice, it was found that the diabetes in mice disappeared and the mice remained diabetes free for months.

3.5 Glucose Control in Critically Ill Patients

Critically ill patients are different from non-critically ill patients in many respects. Major surgery and critical illness are physiologically stressful events that provoke complex metabolic responses in the patient. In general, the greater the degree of surgical trauma, the greater the endocrine upset.

Critical care medicine uses the term "stress" to describe the systemic response to severe injury or infection [7]. Some of the responses to stress include alterations in carbohydrate metabolism, and a state of hypermetabolism. The hypermetabolic state is induced by sepsis or injury, as well as by organs involved in the immunologic response to stress [7]. It is manifested as an increase in glucose production that appears to be directed toward maintaining glucose delivery to wound and immune tissues [7]. These alterations make the control of plasma glucose more difficult than in ordinary diabetic patients.

Factors that affect glucose regulation in critically ill patients include:

- Loss of the inhibitory effects of elevated glucose levels.
- Stress-induced release of counter-regulatory hormones
- Increased Insulin resistance
- Medication that induces hyperglycaemia
- Effects of Feeding
- Gastro-intestinal (GI) tract functionality
- Stress in liver

3.5.1 Loss of the Inhibitory Effects of Elevated Glucose Levels

In the non-stressed state, elevated serum glucose levels exert an inhibitory effect on gluconeogenesis. In critically ill patients, this important feedback mechanism is blunted, often resulting in continued endogenous hepatic glucose production and hyperglycaemia. Although there may be an increase in glucose production and turnover, the reliance on glucose as an energy source is reduced.

3.5.2 Stress-Induced Release of Counter-Regulatory Hormones

In response to stress, the counter-regulatory or anti-insulin hormones are secreted. These hormones include epinephrine, norepinephrine, cortisol, growth hormone, and glucagon.

- Both norepinephrine and epinephrine cause constriction of essentially all blood vessels of the body (vasopressor activity), and an increased activity of the heart (inotropic activity). The constriction of the vessels increases the total peripheral resistance and elevates arterial pressure. Epinephrine (also known as adrenaline) also mobilises glucose from glycogen and raises blood glucose level.

- Cortisol (a gluco-corticoid) stimulates hepatic glycogen and glucose production, and inhibits insulin action on peripheral tissues. It stimulates protein synthesis, providing substrate for gluconeogenesis. Gluco-corticoids also suppress the inflammatory and immune response [4].

- Growth hormone increases the rate of protein synthesis in all cells of the body, and increases mobilisation of fatty acid from adipose tissues into the blood. There is also increased use of the fatty acid for energy, with a decreased rate of glucose utilization throughout the body [4].

In all, the net effect of these hormones is the raising of blood glucose concentration, mobilisation of alternative fuels, reduction in glucose tolerance and increase in peripheral resistance to the effects of insulin.

3.5.3 Increased Insulin Resistance

Patients with Type II diabetes (also know as NIDDM) are usually insulin resistant. Surgical stress potentiates this insulin resistance, mainly due to the release of the counter-regulatory hormones. Enhanced gluconeogenesis during stress is also resistant to inhibition by insulin and glucose. Although skeletal muscle has traditionally been implicated as the major site of peripheral insulin resistance, stress may also induce insulin resistance in adipose tissue, liver, and heart [7].

3.5.4 Medication that Induces Hyperglycaemia

Concurrent administration of exogenous vasopressors and gluco-corticoids can further amplify the effects of stress-induced counter-regulatory hormones in ICU patients as described above. Drugs with vasopressor and/or inotropic activity (such as epinephrine and norepinephrine) are useful for resuscitation of critically ill patients, and are administered when fluid resuscitation alone fails to reverse hypotension.

3.5.5 Effects of Feeding

Patient in Intensive Care are usually sedated. During this period, nutrition is delivered either enterally or parentally.

- Enteral nutrition delivery is always the preferred route whenever the GI tract is functional. Nutrition is delivered proximal or distal to the pylorus through a naso-enteric tube. Since gastric emptying is often impaired in critically ill patients, feeding distal to the pylorus maybe preferable [165].

- Total parenteral nutrition (TPN) is given in situations where the patient has contra-indications to enteral feeding (such as bowel obstruction, overwhelming intra-abdominal sepsis, or nectronizing pancreatitis), or when the patient is unable to absorb nutrients via the GI tract. TPN in the ICU setting is delivered through a central venous catheter, since most critically ill patients already have central venous monitoring in place.

The plasma glucose concentration represents a balance between the influx of glucose into the circulation and the rate at which it is cleared from the circulation (either by the action of insulin, peripheral uptake or due to renal excretion). Taken together with the varying rates of intravenous dextrose-containing fluids ordered by the medical team, the ultimate glucose level in the patient depends on how well this balance is being maintained.

3.5.6 GI Tract Functionality

In terms of meal intake, perhaps the difference between an outpatient and a critically ill patient is in "how" and "when" they receive their daily nutrition.

The direct intubation to the pylorus implies that food ingestion does not go through any peristaltic action of the esophagus, nor the normal stomach stimulation by the presence of food, as would occur in a normal individual. There is the effect of no chewing, no visual stimulation by food, and no activity of the stomach which pre-empts insulin release in sedated patients.

Food absorption in critically ill patients remains dependent on gut motility and other functions of the GI tract. The meal pattern is also abnormal in that nutrient is given continuously over hours rather than as discrete meals.

3.5.7 Stress in Liver

The liver functions as an important blood glucose buffer system. During the blood glucose rise after a meal, the liver stores as much as two thirds of the glucose absorbed from the gut in the form of glycogen [4]. The stored glycogen is later released back into the circulation as glucose when required.

This action of the liver decreases the fluctuations in blood glucose level. However, in patients with severe liver disease, it becomes difficult to maintain a narrow range of blood glucose level [4].

3.6 BG Management in the ICU

Critically ill patients are treated by the method of continuous administration of intravenous insulin by infusion pump, although intermittent subcutaneous injections may also be given in certain cases. Continuous intravenous infusion obviates concerns over the patient's state of perfusion (which influences subcutaneously injected insulin), and permits fine-tuning of BG level within a chosen range:

- 8.3–13.8 mmol/l (150–250 mg/dl) [158]
- 6.7–10 mmol/l (120–180 mg/dl) [8, 166]
- 6–10 mmol/l (108–180 mg/dl) [167].

In the current clinical setting, BG level less than 6 mmol/l (108 mg/dl) carries the potential risk of hypoglycaemia, while BG level in excess of 13.8 mmol/l (250 mg/dl) would also require intervention[1].

Continuous infusion of insulin may increase the risk of hypoglycaemia, but the continuous monitoring of patients in an ICU minimises this risk and enables BG to be controlled safely.

3.6.1 BG Measurement

Current clinical practices examine BG two- to four-hourly, with hourly checks during critical cases. BG determination is by conventional glucometer with the blood sample obtained from arterial or venous cannula.

---

[1] These "higher" ranges would now have to be re-considered in light of the new finding in [9] that emphasized the importance of tighter BG level in critically-ill patients.

3.6.2 Insulin Infusion Adjustment

Insulin infusions are administered in saline or colloid solution, and mixed such that 1ml/hr of delivery equates to 1U/hr of insulin. A rate of 0.5 to 1 unit of insulin per hour is the recommended starting dose. This dose is usually for patients who are not severely stressed. Higher rates may be needed for adequate BG control.

Adjustments to insulin rate are made hourly, two-hourly or four hourly based on arterial blood (from arterial cannula) or fingerstick glucose determinations. The adjustment frequency depends on the stability (or severity) of the BG elevation. Patients may vary greatly in their sensitivity to insulin and some, especially long-term type 1 diabetic patients are quite sensitive to small changes in insulin dose.

Various rules for selecting the appropriate insulin infusion rate exist, such as the sliding scale table, titration, and variable-rate intravenous insulin infusions (see Section 4.2 in Chapter 4). These methods were designed and mainly used for critically ill patients, whose physiological situation is different from ambulatory patients. These method also used intermittent BG determination, rather than continuous measurements.

3.7 Conclusion

Biological systems involve complex and interdependent processes. High blood glucose levels in patients are often a result of metabolic derangement at various levels. Glucose control in critically ill patients differs from diabetic ambulatory patients in that they face additional metabolic stress from critical illness or surgery. This renders their blood glucose level difficult to predict and sometimes control.

3.8 Summary

In this chapter, the glucose-regulatory mechanisms in a normal and diabetic individuals were compared and contrasted. The challenges of controlling the BG levels in the critically ill population were also discussed, followed by an overview of the clinical practices currently used in managing BG in the Intensive Care Unit.

4

Mathematics of Glucose Control

4.1 Introduction

We have looked at blood glucose measurement, insulin infusion (Chapter 2), and the characteristics of the patient in terms of the blood glucose control (Chapter 3). In this chapter, we look at the control algorithms (or the "smarts") that, when worked together with the glucose sensor and insulin infusion pump, would ideally re-balance a patient's blood glucose level.

Recall that the beta cells (which being the fuel sensor as well as the insulin production source) were destroyed in a diabetic patient, causing an impairment of the ability to self-regulate glucose level. Glucose level regulation must then be restored by means of carefully calculated external insulin infusion. The goal of a closed-loop control system is thus to mimic the functionality of the pancreas in providing automatic regulation of blood glucose level in patients.

To be precise, the closed-loop control systems (with its algorithm) should really answer the question: "How much insulin should be given such that the person blood glucose is restored, as closely as possible, to that of a healthy individual?"

Numerous researches were conducted to address this question, and many solutions were proposed. This chapter aims to give a brief introduction to:

1. the control algorithms proposed in the literature, categorised based on the two different approaches to the *design* of closed-loop control algorithm, namely,
   - model-less approach,
   - model-based approach, where the model is linear and/or non-linear.

2. the relationships between the control algorithms and the formulation of glucose-insulin model for the purpose of blood glucose level control in diabetic patients.

This chapter only touches upon some basics of glucose-insulin modelling. Readers interested in detailed mathematical modelling techniques are invited to consult other dedicated textbooks (e.g. [168] and [169]).

F. Chee & T. Fernando: Closed-Loop Control of Blood Glucose, LNCIS 368, pp. 59–108, 2007.
springerlink.com   © Springer-Verlag Berlin Heidelberg 2007

4.2 Model-Less (Empirical) Control Algorithms

In the model-less (empirical) approach to control algorithm design, the relationship between the input (insulin) and output (desired glucose level) are determined based on experimental data, not on a theory. A control rule is then formulated using the experimental data as the basis.

In practice, it is not often easy (or feasible) to completely understand how the body works. Hence, the simplest way to arrive at a control algorithm would be to observe the cause-and-effect (or dose-and-response) in the real system, and formulate a control rule based on the observed relationship between the variables of interest.

4.2.1 Control Algorithm Based on Curve-Fitting

In this method, the relationship between the inputs and outputs are obtained by fitting simple curve equations to the experimental data. As an example, experiments could be conducted to observe the glucose level measured from a patient when different amounts of insulin injections were given over a period of time. A curve would then be fitted to arrive at a simple glucose-insulin response curve (see Table 4.1), which would then be used as the control rule.

The earliest literature featuring the use of glucose-insulin response curve for glucose control was authored by Albisser et al [10, 17, 23, 172].

The control equation formulated by Albisser et al consists of a sigmoidal dose-response curve, with an incorporated predictive equation.

The predictive portion of Albisser et al's algorithm takes into account:

- the trend of BG (i.e. rise, fall)
- the delay between blood extraction and the ultimate measurement

The control equation is likened to a Proportional-Derivative controller. The derivative action is based upon the rise or fall of BG, while the dose-response curve provides the Proportional action.

In those time, the patient's BG was measured invasively on a minute-by-minute basis (see Section 2.3), and insulin was infused intravenously. However, a delay of 5–10 min were required before a BG reading was available due to the then-available BG analysis method. It was reported that any delay in the BG measurement could throw the controller action out of synchronisation [22]. Compensation was required by predicting BG based on the rates of change over the previous minutes.

Response to a meal was usually normalised by the combined action of a rapid insulin delivery on first detection of BG rise (i.e. the derivative action of the controller), and the action of a second phase secretion performed by the algorithm. The timeliness of the first phase insulin release was compensated for, through minute-by-minute BG measurement, and the fact that insulin entered the circulation much more rapidly when administered intravenously (compared to subcutaneous infusion) [156].

Unlike other methods that relied on variables that correlated with the whole blood glucose, invasive BG measurement has the advantage of being more accurate, because it measured glucose content in the whole blood itself.

Table 4.1. Glucose-insulin response curve (Sigmoidal dose-response curve)

Glucose-Insulin Response Curve (Sigmoidal Dose-Response Curve)

The glucose-insulin response curve came from the discovery that insulin secretion did not respond as a linear function of glucose concentration. The relationship between the extracellular glucose concentration and the rate of insulin secretion in vitro is sigmoidal, with a threshold corresponding to the glucose level normally seen under fasting conditions, and with the steep portion of the dose-response curve corresponding to the range of glucose levels normally achieved postprandially.

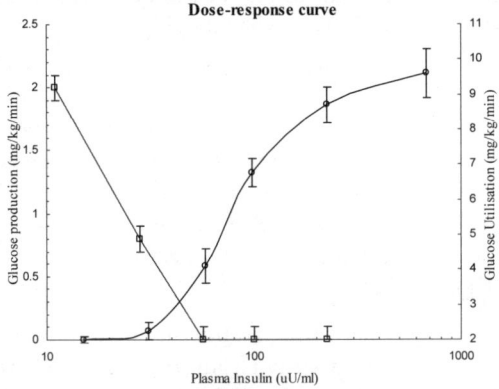

This sigmoidal nature of the dose-response curve could be attributed to a Gaussian distribution of threshold for stimulation in beta-cells. This also means that linear model is not adequate to describe the glucose-insulin interaction in human (Text adapted from [159, 170]; Graph adapted from [171]).

Albisser's algorithm formed the basis for the algorithm used in the Biostator® bedside glucose control system (Table 4.3) used in further clinical investigations of glucose-insulin interaction in humans (e.g. see [17], [16], [173] and [174]).

4.2.2 Control Algorithm Based on Lookup Table

In this method, the relationship between the inputs and outputs is obtained by mapping the inputs and outputs in the form of a lookup table. In the day-to-day care of patients in the hospital settings, blood glucose control is commonly achieved using model-less approach in an open-loop manner. BG samples are taken intermittently (at defined internals), and insulin delivery rate is adjusted manually using:

- lookup table control, such as a sliding scale table. The table has a continuous BG "partitioned" into ranges, with an insulin rate assigned to each range.

Table 4.2. Albisser et al's infusion algorithm

Albisser et al's Infusion Algorithm

- Glucose infusion rate, $R_d = \frac{1}{2} M_d \left[1 - \tanh S_d \left(G - B_d\right)\right]$
- Insulin infusion rate, $R_i = \frac{1}{2} M_i \left[1 - \tanh S_i \left(G_p - B_i\right)\right]$
- Projected BG, $G_p = G + K_1 \left[\exp\left(\frac{A}{K_2}\right) - 1\right]$

where M = maximum infusion rate; S = slope; B = BG level at which half maximum infusion rate is chosen to occur; subscript d,i = for glucose/dextrose and insulin respectively; A = rate of change of BG, which is minute-to-minute changes of BG averaged over the preceding four minutes; K_1 is chosen to adjust the magnitude of the difference factor, and K_2 selected to establish its sensitivity to changes in A.

Insulin is given in accordance with the range in which the BG sample resides (Table 4.4). The insulin delivery can either be intravenous or subcutaneous, and different tables are prescribed for different routes of delivery. Tables that used 1-hourly [175], and 3-hourly BG sampling [18] for subsequent adjustment of intravenous insulin delivery rate have been reported to achieve good normalisation of high BG.

- linearised lookup table, where the "step-wise" insulin increase is replaced by a "slope". Furler et al [176] used such an algorithm, where insulin rates are 0.5 U/hr for BG<4 mmol/l and 2.5 U hr for BG > 8 mmol/l, with a linear transition between these rates over the range of BG of 4–8 mmol/l. Ollerton [177] subsequently improved on Furler et al's algorithm by using grid search algorithm. These slope-based algorithms were mainly used in clinical investigation (e.g. [176]).

Nowadays, experienced nursing staff would generally be able to control an inpatient's elevated glucose level more efficiently than a prescribed sliding table. This is because a prescribed sliding table can lose it effectiveness when the patients glucose tolerance changes with their state of well-being.

4.2.3 Control Algorithm Based on Rule-Based Control

Model-less approach to blood glucose control could also use expert rules (rules that are based on experience) as the control rules. They can be seen in clinical techniques, which include:

- titration control, where the intravenous (IV) insulin infusion rate is started at an empirical level, and progressively tuned to an appropriate rate which lowered and maintained the BG in the target range;

- variable-rate IV insulin infusion algorithm, where insulin rates are increased/decreased by 0.5 U/hr (or remain unchanged) every 2 hours, based

Table 4.3. Biostator® algorithm at a glance

Biostator® algorithm

Throughout the evolution of Biostator®, a series of control algorithms were developed. These newer algorithms addresses the various observations encountered in their predecessor algorithms, and attempts to fine-tune the BG response. Broekhuyse et al [175] compared the various algorithms used in Biostator® (or so called "GCIIS - Glucose-controlled Insulin Infusion System"). The original Albisser's algorithm used both exogenous glucose and exogenous insulin infusion in the control action. Later, modifications in algorithms permitted a blunting of insulin delivery when glucose concentrations decline. This has eliminated the requirement of such a glucose counter-infusion. The general structure of the earlier algorithm has the form:

$$I = \frac{1}{2} M_I \left[1 + \tanh S_I \left(G_p - B_I \right) \right]$$

where I is the insulin infusion rate [mU/min], M_I is the maximum insulin infusion rate, S_I is the sigmoidal steepness factor, B_I is the glucose concentration at the half-maximum insulin infusion rate, G_p is a calculated quantity called "projected glucose":

$$G_p = \begin{cases} G_0 + K_1 \left[\exp \left(\frac{A}{K_2} \right) - 1 \right] & \text{(Albisser)} \\ G_0 + \left(K_1 \cdot A^3 + K_2 \cdot A \right), & A = \left(4G_0 - G_1 - G_2 - G_3 \right)/10 \text{ (Toronto)} \\ G_0 + K_1 \left(\exp \frac{RC}{K_2 - 1} \right), & RC = \frac{1}{15} \sum_{i=1}^{i=4} 2^{4-i} \left(\frac{\Delta G}{\Delta t} \right)_{t-i} \text{ (Kraegen)} \end{cases}$$

where A is the linearly weighted average rate of change of glucose concentration based on the previous four minutes of monitoring; $G_0 =$ last minute's average glycaemia [mg/dl]; $G_k =$ the average glycaemia $k+1$ minute previously [mg/dl] for $k = 1, 2, 3, 4$; K_1 and K_2 are constants (see [175]). Biostator® algorithm later introduced three different settings [16, 22, 176, 177]:

- Static: insulin release is dependent upon the static value of BG:

$$IR = R_i \cdot \left(\frac{G_0 - B_I}{Q_i} + 1 \right)^n$$

- Dynamic: insulin infusion is solely controlled by the rates of change of BG:

$$IR = R_i \cdot \left(\frac{G_0 \mid G_D \quad B_I}{Q_i} + 1 \right)^n$$

- Dynamic+Static: the static with the dynamic control mode for insulin infusion:

$$IR = R_i \cdot \left(\frac{G_0 + G_D - B_I}{Q_i} + 1 \right)^n - R_i \cdot \left(\frac{G_0 - B_I}{Q_i} + 1 \right)^n$$

where $G_D = \begin{cases} \frac{K_R A^2}{10} + 6A \text{ for } A > 0 \\ \frac{K_F A^2}{10} + 6A \text{ for } A < 0 \end{cases}$ and $n = 2$ or 4 depending on how much insulin release is required [176].

The development of the control algorithm saw the use of K_R when BG is increasing and K_F when BG is decreasing ($K_R > K_F$), instead of maintaining a fixed control constant for both the increase and decrease in BG.

Table 4.4. Example sliding scale table

BG range (mmol/l)	Insulin infusion rate (U/hr)
>20.0	4
15.1 − 20.0	3
10.1 − 15.0	2
6.1 − 10.0	1
0 − 6.0	0

on the BG measurement. This method is similar to a sliding table, except that the insulin rates are given in terms of increment or decrement to the previous rate, instead of a fixed insulin rate for each BG ranges. Proposed by Watts et al [8] to maintain BG in patients within a certain range, this method was reported to be safe and efficacious [2, 8];

- experience, where control is achieved by means of accumulated experiences in controlling glucose level;

- neuro-fuzzy method, where insulin rates are increased/decreased every 4 hours, based on a simple nomogram (similar to a look-up table). The nomogram details the insulin infusion rate variation to be added or subtracted from the current rate based on "preceeding" and "present" BG values. The nomogram is obtained by training a Back Error Propagation Neural Network using 1000 paired BG–insulin values, and then providing the neural network with 400 pairs of BG values that represents every possible combination of glycaemic values between 3.3 and 13.9 mmol/l to obtain 400 corresponding variations of the insulin infusion rates [166].

Note that the glucose-insulin response curve (and other algorithms such as Furler et al's infusion slope) can be discretised (or partitioned) to form a sliding table. How closely the discretised sliding table approximates the original curve depends on the selection of the BG ranges and the accuracy of the conventional infusion pump (i.e. what is the smallest reliable infusion rate available).

4.2.4 Control Algorithm Based on PID Control

Proportional-Integral-Derivative control (or any combination of P, I or D) is an easy-to-use feedback control system. It does not require advanced mathematics to design and can be easily "tuned". The controller takes a measurement from a plant process and compares it with a setpoint (reference) value. The difference (or "error" signal) is then used to adjust the input to the plant in order to bring the measured value back to its desired setpoint (Fig 4.1). PID controller can adjust process outputs based on the history and rate of change of the error signal. However, PID controller are sensitive to dead-time in the measurement (i.e. delay in measurement could upset the control action, bringing the control loop into possible oscillations).

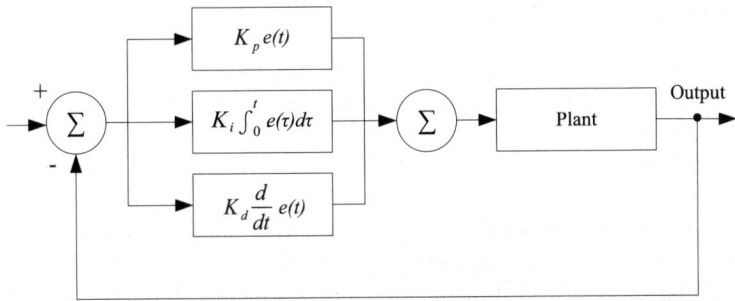

Fig. 4.1. PID controller

- Proportional (P): The error signal is multiplied by a constant K_p for immediate correction.

- Integral (I): To learn from the past, the error signal is integrated (added up) over a period of time, and then multiplied by a constant K_i. The result is then added to the controller output signal.

- Derivative (D): The slope of the error signal (i.e. the change of the error signal over a pre-defined interval) is calculated, and multiplied by a constant K_d, and the the result is added to the controller output signal. The larger the derivative term, the more rapidly the controller response to changes in the plant output.

PID controller can be divided into two forms:

1. "Position" Form

   In Position form, the new value of the controller output is calculated directly, i.e.:

   $$u(t) = K_p e(t) + K_i \cdot \Delta t \left[\sum_{i=1}^{n} e(i) \right] + K_d \frac{[e(t) - e(t - \Delta t)]}{\Delta t}$$

   where Δt = time between data samples (which must be smaller than the process response time).

2. "Velocity" Form

   In Velocity form, only the change in the controller output is calculated. The actual controller output is then calculated from the previous value, i.e.:

   $$\Delta u = u(t) - u(t - \Delta t)$$
   $$= \underbrace{K_p \left[e(t) - e(t - \Delta t) \right]}_{P} + \underbrace{K_i \Delta t \cdot e(t)}_{I}$$
   $$+ \underbrace{K_d \frac{[e(t) - 2e(t - \Delta t) + e(t - 2 \cdot \Delta t)]}{\Delta t}}_{D}$$

Implementations in Literature

- Palazzo et al [178] simulated a PID controller on Bergman's model.
- Chee et al [167, 179] implemented a step-wise PID controller on critically-ill patient population.
- Steil et al [180] implemented a PID controller, which is subsequently used in a "Long Term Sensor System, LTSS" system (MiniMed-Medtronic, Northridge, CA, USA) (see [181]). LTSS combines an intravenous glucose sensor and an implanted insulin pump that uses intraportal infusion route).
- Gupakumaram et al [182] implemented a P-D controller that has a feed-forward component. This component models the pancreatic beta-cell response in addition to present and previous glucose values. The beta-cell insulin secretion is modelled on a fading history of glucose levels and trends, where current and recent glucose levels are weighted more heavily than older glucose values. The recent trends in glucose variations are also weighted more heavily than those in the past:

$$I(t) = K_0 \int_0^t W_0(t) \cdot G(t) dt + K_1 \int_0^t W_1(t) \dot{G}(t) dt$$

$$+ K_2 \int_0^t W_2(t) \ddot{G}(t) dt + \text{higher order terms}$$

where $I(t)$ = insulin infusion rate; K_0, K_1, K_2 = Gain factors; W_0, W_1, W_2 = weighting functions, and $G(t)$ = blood glucose levels.

- Marchetti et al [183] simulated a PID controller in the velocity form on Hovorka's model. The PID controller has the form

$$c(t) = c_0 + K_c \left(e(t) + \frac{1}{\tau_I} \int_0^t e(t) dt + \tau_D \frac{de(t)}{dt} \right)$$

where K_c and τ_D are tuned to minimise the objective function

$$J = \sum \left[\left(\sum_k^{t_\infty} |e(k)| \Delta t \right) + w_i \right] + c_1 + c_2$$

with $w_i = \exp\left(-100(G_{min,i} - 60)\right)$ being a constraint to avoid hypoglycemia; $c_1 = \exp\left(-10000(-0.01 - K_c)\right)$ being a constraint to avoid K_c values larger than -0.01 mU dl/mg min, and $c_2 = \exp\left(-10000(1500 - \tau_D)\right)$ being the constraint to avoid τ_D values larger than 1500 min. The index $i = c, u, o$ refers to the response for a correct, under-estimated and over-estimated carbohydrate content of the meal; $e(t)$ = control error (mg/dl); t_∞ = duration of the simulation, and Δt = sampling period.

Advantages and Disadvantages of Model-Less Approach

The advantages with model-less approach is that:

1. It is a simple and easy way to relate the input and output variables.

2. Compared to model-based approach, the patient parameters do not need to be worked out in advance, and the control algorithm can be deployed instantly.

Nevertheless, as this approach is empirical in nature:

1. It assumes no knowledge of the underlying system and is likened to the Black box approach.
2. It provides no insight into the functioning of the underlying system.
3. Its predictive power is very much restricted.
4. Successful operation of control algorithm still relies on human expertise to some degree, for example, in the adjustment of the starting insulin rate (as in [8]), or adaptation for use in patients with differing sensitivity to insulin.

4.3 Model-Based Control Algorithms

As the name implies, the model-based approach involves the use of a model in the control of blood glucose level. This model is the human glucose-insulin interaction. If this complex interaction can be captured and described in terms of mathematics, then the glucose control problem becomes a mathematical problem, and mathematical problem can be solved using various mathematical techniques (Note: We are yet to discover that it is not as easy as once thought to model the glucose-insulin interaction).

Other advantages of using a model:

- It offers useful description and insight into the underlying process. For example, the observations obtained from accessible variables (e.g. glucose at tissue) can be used to measure those system quantities that are of interest, but were inaccessible to direct measurement (e.g. glucose at hepatic artery).

- It can help in predicting overall system behaviour under a variety of perturbations [168]. In another word, it can serve to determine how a system would respond to a stimulus or changes in the system.

- It allows the testing or simulation of the control algorithm to be performed without involving real patients.

- It allows the study into the effect of insulin therapy regime to be undertaken without risking patient safety.

Nevertheless, we must understand that whether or not a certain model is valid in terms of its representation (approximation) of a real system really depends on the context (or purpose) to which it is designed for.

Furthermore,

- Lack of understanding about the underlying process limited the extent of derivation of the model.

- The predictive power of the model is limited by the extent to which the model is valid or accurate.

- Some parameters in the model may not be always observable, and if they cannot be measured in anyway (or are too difficult to be measured), then this will limit the usefulness and applicability of the model.

The mathematical model used in BG regulation can be divided into two groups:

4.3.1 Theoretical

Theoretical modelling attempts to model the underlying process by means of a theory. The theory is derived using the knowledge about the function or structure of the underlying system, chemical process, physical laws, and quantitative data (from observable measurements).

The extent to which a pure theoretical model can be derived depends on current understanding of the system or process. It may not always be possible to fully model the underlying process, either due to a lack of measurable variables, or simply lack of detailed knowledge about the system.

4.3.2 Empirical + Theoretical

In metabolic system, it is rare to have adequate detailed knowledge of the underlying system to allow full derivation of the theoretical models.

Hence, a combination of theoretical and empirical knowledge is used. Certain parts of the system are modelled empirically, whereas others are modelled theoretically.

We could postulate about the internal structure of a system based on empirical observation, and fit the hypothesis to the observation.

When deriving the glucose-insulin model, one could conceptualise the model based on

- physiological knowledge of the system,
- functional description of the relevant processes
- the interconnection between the processes, and
- Relationship between the processes to the observable glucose (or insulin) measurements in practical situation

Certain elements of the model could also be, as according to [168]:

- Abstractised, i.e. only certain aspects of the blood glucose regulation system are considered in a model. For example, bloodstream is regarded as containing only glucose and hormones involved in its regulation. Features that do not directly relate to glucose regulation are omitted.
- Aggregated, i.e. lumping different components into a single entity for the purpose of modelling. For example, the tissues that constitute the glucose utilization space are normally lumped together, thus ignoring individual cellular properties.

• Idealised, i.e. some structures or behaviours that are difficult to describe or treat are approximated by more simple idealized ones. For example, insulin injected directly to the bloodstream takes effect instantaneously although in fact its action is delayed.

Before continuing with a discussion of model-based control algorithms, it is useful to have some knowledge of the mathematical models that were used in the design of the algorithms.

4.4 Mathematical Models of Gluco-regulatory System

The attempts to capture the glucose-insulin mechanism have resulted in the formulation of various glucose-insulin kinetic models. These models range from simple expressions that relate glucose and insulin, to very complete mathematical models. The three general groupings of mathematical models are:

1. Linear (see e.g. Bolie et al [184], Ackerman et al [15, 185–187], Ceresa et al [188], Chorbajian et al [189], Cerasi et al [190], Nomura et al [191], Bajaj et al [192,193], Jansson et al [194] and Salzsieder et al [195])
2. Non-linear (see e.g. Bergman et al [196–198], Furler et al [176], Candas & Radziuk et al [199], Doyle III et al [200] and De Gaetano et al [201])
3. Comprehensive (see e.g. Cobelli et al [202–204], Parrish et al [205], Sorensen [206–209], Hovorka et al [210,211], Giugliano et al [212], and Guyton et al [4]).

4.4.1 Linear Models

Linear models are adequate when the intrinsic dynamics of the metabolic system are essentially linear[1]. In linear modelling of glucose-insulin kinetics, the models are described by linear time-invariant equations:

$$\dot{\mathbf{x}}(t) = A\mathbf{x}(t) + B\mathbf{u}(t)$$
$$y(t) = C\mathbf{x}(t) + D\mathbf{u}(t); \qquad t_0 < t < T \qquad (4.1)$$

where the state variable $\mathbf{x}(t)$ and its derivative appear in linear combination only, and $\mathbf{u}(t)$ represents the input (or disturbances) into the system. Models of glucose-insulin kinetics can be derived using technique like "compartmental analysis".

---

[1] Quoted from [213]: A system is called "linear" if the Principle of Superposition applies. The Principle of Superposition states that the response produced by a simultaneous application of two different forcing functions is the sum of the two individual responses. Hence, for the linear system, the response to several inputs can be calculated by treating one input at a time and adding the results.

Conversely, a system is "nonlinear" if the Principle of Superposition does not apply. Thus, for a nonlinear system the response to two inputs cannot be calculated by treating one input at a time and adding the results.

Compartmental Analysis (Adapted from [168])

Compartmental analysis is the simplest method in biomathematics to describe the transfer of materials in biological systems, and it can quickly lead to mathematical relationships.

A compartment is fundamentally an idealised store of a substance. Compartmental analysis consists of studying the exchanges of matter between the stores (i.e. compartments) as a function of time, t. The material exchange between compartments takes place either by physical transport from one location to another, or by chemical reactions (Fig 4.2). The mathematical model then consists of the mass balance equations for each compartments and relations describing the rate of material transfer between compartment:

$$\frac{dQ_{ij}}{dt} = \sum R_{ij} \sum R_{ji}$$

where Q_{ij} = quantity of substance in compartment i that interchanges matter with other compartments; $\sum R_{ij}$ = summation of the rates of mass transfer into compartment i from all relevant compartments; $\sum R_{ji}$ = summation of the rates of mass transfer from compartment i to other compartments of the system.

For example, consider two compartments 1 and 2. Figure 4.2 shows the flow of material between the two compartments. K_{ij} indicates the rates at which the materials in i is transferred to compartment j and vice versa. Let Q_1, Q_2

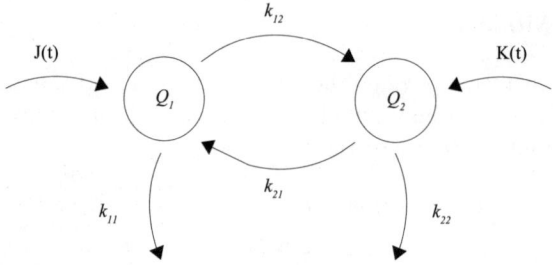

Fig. 4.2. Compartmental analysis

= quantity of materials in compartment 1 and 2 respectively, and $J(t)$, $K(t)$ = flow of material from exogenous sources, the mass balance equations can be written as follows:

$$\frac{dQ_1}{dt} = -k_{11}Q_1 - k_{12}Q_1 + k_{21}Q_2 + J(t)$$

$$\frac{dQ_2}{dt} = k_{12}Q_1 - k_{21}Q_2 - k_{22}Q_2 + K(t)$$

which can be simplified to:

$$\frac{dQ_1}{dt} = -m_1Q_1 + k_{21}Q_2 + J(t)$$

$$\frac{dQ_2}{dt} = k_{12}Q_1 - m_3Q_2 + K(t) \qquad (4.2)$$

where $m_1 = (k_{11} + k_{12})$ and $m_3 = (k_{21} + k_{22})$.

The form taken by equation 4.2 above is the most common form of models for linearised model of glucose-insulin, where the equation consist of glucose and insulin as the primary compartments.

4.4.2 Ackerman's Linear Model

There are many linear models proposed in the literature to-date. Among them, Ackerman's model has been by far the most commonly cited linear model. Ackerman's model consists of a system of equations in which the parameters have been lumped into two dependent variables, G and H, and four rate constants:

$$\frac{dg}{dt} = -m_1g - m_2h + J$$

$$\frac{dh}{dt} = -m_3h + m_4g + K \qquad (4.3)$$

where $G=$ glucose concentration; $G_0=$ fasting glucose concentration; $g \equiv G-G_0$; $H=$ blood hormone concentration (including insulin). This is the effective hormone concentration which included consideration of the role of epinephrine and other hormone; $H_0=$ fasting blood hormone concentration (including insulin); $h \equiv H - H_0$

The rate constants are:

- $m_1=$ rate constant for the removal of glucose above the initial fasting level due to its own excess above the initial level. Also known as glucose effectiveness, this term normally has a value of 0.01–0.02/min but it is likely that its true value is reduced at chronic hyperglycemia due to glucose toxicity;
- $m_2=$ rate constant for the removal of glucose above the initial level due to blood hormone concentration above the initial level
- $m_3=$ rate constant for the removal of hormone above the initial fasting level due to its own excess above the initial level (duration of insulin action)
- $m_4=$ rate constant for the removal of hormone above the initial level due to blood glucose concentration above the initial level.

Note the similarities between equation 4.3 and the compartmental model (equation 4.2)). The same sets of equation 4.3 can also be derived using compartmental analysis.

The model parameters for the model were obtained from fitting the glucose and insulin concentration profile to the model, following glucose tolerance tests in patients. Some typical values can be found in Appendix B.

Table 4.5. Bolie vs Ackerman

Bolie's Model

In 1961, Bolie published his pioneering work on mathematical modelling of the glucose-insulin interaction, which was presented to show that "the normally functioning glucose homeostasis mechanism exhibits physiological coefficients which approximate the well-known critical damping criteria of servo-mechanism theory" [184].

The model is represented by the equations:

$$\frac{dH}{dt} = \frac{1}{V}\dot{U} - \frac{F_1(H)}{V} + \frac{F_2(G)}{V}$$

$$\frac{dG}{dt} = \frac{1}{V}\dot{P} - \frac{F_3(G,H)}{V} - \frac{F_4(G,H)}{V}$$

where t = time [hours]; G = extracellular glucose concentration; H = extracellular insulin concentration; V = extracellular fluid volume [L]; $\dot{U}$ = rate of insulin injection [U/hr]; $\dot{P}$= rate of glucose injection [g/hr]; $F_1(H)$ = rate of insulin destruction; $F_2(G)$ = rate of insulin production; $F_3(G,H)$ = rate of liver accumulation of glucose; $F_4(G,H)$ = rate of tissue utilization of glucose.

Assume that the fluctuations in the insulin and glucose concentrations are limited to small physiological variations, then the non-linear model can be linearised using Taylor series:

$$y = f(x)$$

$$= f(\bar{x}) + \frac{df}{dx}(x - \bar{x}) + \frac{1}{2!}\frac{d^2f}{dx^2}(x - \bar{x})^2 + \ldots \qquad (4.4)$$

to produce

$$\frac{dh}{dt} = u - \left[\frac{1}{V}\left(\frac{\partial F_1}{\partial H}\right)_{H_0}\right]\cdot h + \left[\frac{1}{V}\left(\frac{\partial F_2}{\partial G}\right)_{G_0}\right]\cdot g$$

$$\frac{dg}{dt} = p - \left[\frac{1}{V}\left(\frac{\partial F_3}{\partial H} + \frac{\partial F_4}{\partial H}\right)_{H_0,G_0}\right]\cdot h - \left[\frac{1}{V}\left(\frac{\partial F_3}{\partial G} + \frac{\partial F_4}{\partial G}\right)_{H_0,G_0}\right]\cdot g$$

where $h = H - H_0$, $g = G - G_0$, $u = \frac{1}{V}\dot{U}$, and $p = \frac{1}{V}\dot{P}$.

The equation above can be identified with Ackerman's model in terms of the parameters m_1, m_2, m_3 and m_4.

4.4.3 Non-linear and Comprehensive Models

The linear model has the disadvantage that it is a gross oversimplification of the underlying glucose-insulin interaction in actual human (which is far more complex than the linear model). Biological system dynamics are often non-linear in nature, and low-order models may not adequately describe the real process and therefore could contain both unacceptable levels of modelling error and significant process-model mismatch [200].

Non-linear model ranges from less complex ones (e.g. [196, 197], [176] and [214]) to comprehensive ones (e.g. [202–204], [215, 216] and [217]). Comprehensive model attempts to coalesce the knowledge of metabolic regulations into a generally large, non-linear model of high order, with a large number of model parameters. This included the modelling of the distribution and metabolism of glucose and insulin, hepatic glucose balance (i.e. glucose production and disposal), renal excretion, glucose utilization, and insulin release and degradation, to describe the system thoroughly ([202–204]). Some investigators even took the molecular approach, modelling individual isolated beta-cells, followed by populations of beta-cells (see e.g. [212]). Comprehensive models, in general, cannot be easily identified.

In this section, only a few of the nonlinear models commonly referred in the literature are described.

4.4.4 Minimal Model

Also known as Bergman's model, the minimal model was proposed by Bergman et al [196, 197] in the early 1980's for the interpretation of glucose and insulin plasma concentrations following the intravenous glucose tolerance test (IVGTT). This model has been popular among past physiological researches on the metabolism of glucose.

It is composed of two parts:

1. **Minimal model for glucose disappearance:** Two differential equations which describe the glucose plasma concentration time-course, accounting for the dynamics of glucose uptake dependent on, and independent of circulating insulin. It treated insulin plasma concentration as a known forcing function [201]

$$\frac{dG(t)}{dt} = -\left(p_1 + X(t)\right)G(t) + p_1 G_B \tag{4.5}$$

$$\frac{dX(t)}{dt} = -p_2 X(t) + p_3[I(t) - I_B] \tag{4.6}$$

where the term $p_1 G_B$ accounts for the body's natural tendency to move towards basal glucose levels [177].

2. **Minimal model for insulin kinetic:** Single equation which describes the time-course of plasma insulin concentration, accounting for the dynamics of pancreatic insulin release in response to the glucose stimulus. The glucose plasma concentration is to be regarded as a known forcing function.

$$\frac{dI(t)}{dt} = \begin{cases} \gamma[G(t) - h]t - n[I(t) - I_B] & for\ G(t) - h > 0 \\ -n[I(t) - I_B] & for\ G(t) - h \leq 0 \end{cases} \tag{4.7}$$

where the state variables:

- $G(t)$ [mg/dl] = the blood glucose concentration at time t [min];
- $I(t)$ [μU/ml] = blood insulin concentration at time t [min];
- $X(t)$ [min^{-1}] = a function representing insulin-excitable tissue glucose uptake activity, proportional to insulin concentration in a "distant" compartment;
- G_B [mg/dl] = the subject's basal glucose level (or pre-injection value of glucose concentration);
- I_B [μU/ml] = the subject's basal insulin level (or pre-injection value of insulin concentration);
- G_o [mg/dl] = the theoretical glucose level at time 0 after the instantaneous glucose bolus, i.e. glucose concentration that would be obtained immediately after glucose injection if there were instantaneous mixing of glucose in the extracellular fluid compartment [218].

The parameters are:

- γ [(μU/ml)/(mg/dl)$^{-1}$min^{-1}] = the rate of pancreatic release of insulin after the bolus, per minute and per mg/dl of glucose concentration above the "target" glycaemia;
- h [mg/dl] = the pancreatic "target glycaemia" [201]. It is a threshold value such that only glucose level above h will effect a secretion of insulin [219]. It represents the critical value of plasma glucose at which glucose begins to have a marked influence on the magnitude of the second phase;
- n [min^{-1}] = the time constant for insulin disappearance [197], or fractional disappearance rate constant for endogenous insulin [198];
- I_o [μU/ml] = the theoretical plasma insulin concentration at time 0, above basal insulinemia, immediately after the glucose bolus.

The two parts are to be separately estimated on the available data. That is, the model parameter fitting has to be done in two steps:

1. using the recorded insulin concentration as input data, in order to derive the parameters in the first two equations
2. then, using the recorded glucose as input data to derive the parameters in the third equation.

The glucose and insulin recordings are obtained by first injecting a bolus of glucose into the bloodstream of the experimental subject, to induce an (impulsive) increase in the plasma glucose concentration $G(t)$ and a corresponding increase of the plasma insulin concentration $I(t)$, secreted by pancreas. These concentrations are measured during a three-hour time interval beginning at injection, until $G(t)$ and $I(t)$ have returned to normal [201].

To represent a person in a diabetic state, the original model has taken the general form:

$$\frac{dG(t)}{dt} = -\left(p_1 + X(t)\right) G(t) + p_1 G_B + p(t)$$

$$\frac{dX(t)}{dt} = -p_2 X(t) + p_3[I(t) - I_B]$$

$$\frac{dI(t)}{dt} = -n[I(t) - I_B] + u(t)$$

where endogenous insulin secretion (i.e. the term $\gamma[G(t) - h]t$) in equation 4.7 was removed, and a term of exogenous infusion of glucose $p(t)$ and insulin $u(t)$ was added.

There were variations to the Minimal Model, for example, Furler et al [176] adapted the original Minimal model to represent the diabetic state and included insulin antibodies in the description of insulin dynamics; Van Herpe et al [220] modified Minimal model for the Intensive Care Unit (ICU) population (called "ICU-MM" model by the authors).

4.4.5 Cobelli's Model

Cobelli et al's model consists of a metabolic plant (glucose), and two-hormone controller (insulin and glucagon) [202–204]. The glucose subsystem is described by a one-compartment model of distribution and metabolism (extracellular fluids), involving net hepatic glucose balance (i.e. the difference between liver glucose production and liver uptake), renal excretion of glucose, insulin-dependent glucose utilization (mainly by muscle and adipose tissue), insulin-independent glucose utilization (mainly by the central nervous system and the red blood cell).

Glucose Subsystem

$$\dot{x}_1(t) = \text{NHGB}(x_1, u_{12}, u_2) - F_3(x_1) - F_4(x_1, u_{13}) - F_5 + I_x(t), \qquad x_1(0) = x_{10}$$
$$\dot{u}_{1p}(t) = -k_{21} u_{1p} + k_{12} u_{2p} + W(x_1), \qquad u_{1p}(0) = u_{1p0}$$
$$\dot{u}_{2p}(t) = k_{21} u_{1p} - (k_{12} + k_{02}(x_1)) u_{2p}, \qquad u_{2p}(0) = u_{2p0}$$

where $\text{NHGB} = F_1(x_1, u_{12}, u_2) - F_2(x_1, u_{12})$ is the net hepatic glucose balance; F_1= the liver glucose production; F_2 = the liver glucose uptake; F_3 = the renal excretion; F_4 = the peripheral insulin-dependent glucose utilization; F_5 = the peripheral insulin-independent glucose utilization; $I_x(t)$ is the rate of exogenous glucose given intravenously; $W(x_1)$ = insulin synthesis controlled by BG concentration; x_1 = quantity of glucose in plasma and extracellular fluids [mg]; u_{1p} = quantity of pancreatic stored insulin [μU]; u_{2p} = quantity of pancreatic, promptly releasable insulin [μU]. The constants are k_{12}=0.01, k_{21}=4.34$\times 10^{-3}$ (values for a normal state).

• Liver glucose production

$$F_1(x_1, u_{12}, u_2) = a_{11} G_1(u_2) H_1(u_{12}) M_1(x_1)$$

where

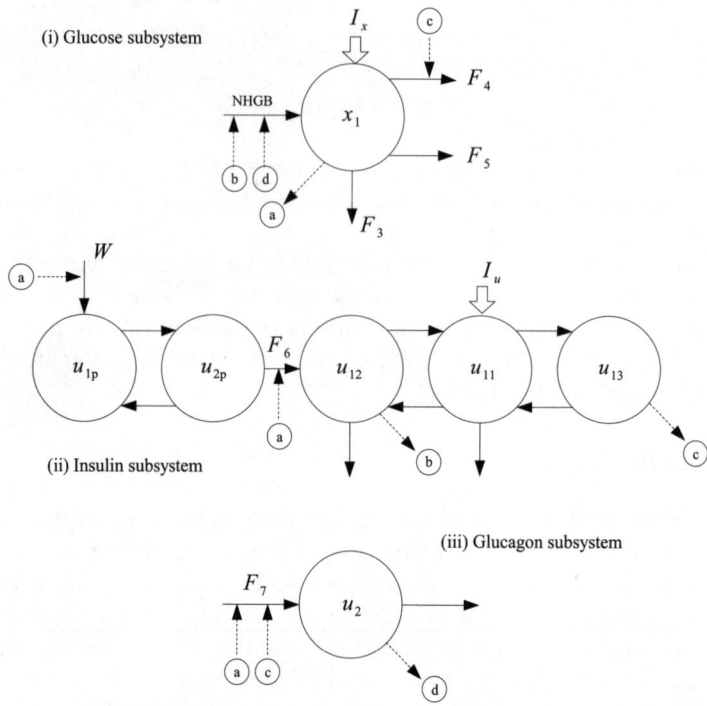

Fig. 4.3. Cobelli et al's compartmental model. Solid line represents material flow; Dashed line represents control signal (Adapted from Cobelli et al [202]).

$$G_1(u_2) = 0.5\left\{1 + \tanh\left[b_{11}\left(e_{21} + c_{11}\right)\right]\right\}$$
$$H_1(u_{12}) = 0.5\left\{1 - \tanh\left[b_{12}\left(e_{12} + c_{12}\right)\right]\right\}$$
$$M_1(x_1) = 0.5\left\{1 - \tanh\left[b_{13}\left(e_x + c_{13}\right)\right]\right\}$$

Baseline parameter values for a normal state: a_{11}= 1.51; b_{11}= 2.14; b_{12} = 0.0728; b_{13}=0.0275; c_{11}=−0.85; c_{12}=7; c_{13}=20.

- Liver glucose uptake
$$F_2(x_1, u_{12}) = H_2(u_{12})M_2(x_1)$$
where

$$H_2(u_{12}) = 0.5\left\{1 - \tanh\left[b_{21}\left(e_{12} + c_{12}\right)\right]\right\}$$
$$M_2(x_1) = a_{221} + a_{222}0.5\left\{1 + \tanh\left[b_{22}\left(e_x + c_{22}\right)\right]\right\}$$

Baseline parameter values for a normal state: a_{221}= 0.00195; a_{222}= 0.00521; b_{21} = 0.0111; b_{22}=0.0145; c_{12}=51.3; c_{22}=−108.5.

- Renal excretion of glucose

$$F_3(x_1) = M_{31}(x_1)M_{32}(x_1)$$

where

$$M_{31}(x_1) = 0.5 \{1 + \tanh [b_{31} (y_1 + c_{31})]\}$$
$$M_{32}(x_1) = a_{321}y_1 + a_{322}$$

Baseline parameter values for a normal state: $a_{321} = 1.43 \times 10^{-5}$; $a_{322} = -1.31 \times 10^{-5}$; $b_{31} = 20$; $c_{31} = -180$.

- Insulin-dependent peripheral glucose utilization

$$F_4(x_1, u_{13}) = a_{41} H_4(u_{13}) M_4(x_1)$$

where

$$H_4(u_{13}) = 0.5 \{1 + \tanh [b_{41} (e_{13} + c_{41})]\}$$
$$M_4(x_1) = 0.5 \{1 + \tanh [b_{42} (e_x + c_{42})]\}$$

Baseline parameter values for a normal state: $a_{41} = 0.0287$; $b_{41} = 0.031$; $b_{42} = 0.0144$; $c_{41} = -50.9$; $c_{42} = -20.2$.

- Insulin-independent glucose uptake

$$F_5(x_1) = M_{51}(x_1) M_{52}(x_1)$$

where

$$M_{51}(x_1) = a_{51} \tanh [b_{51} (e_x + c_{51})]$$
$$M_{52}(x_1) = a_{52}e_x + b_{52}$$

Baseline parameter values for a normal state: $a_{51} = 1.01 \times 10^{-3}$; $a_{52} = 4.6 \times 10^{-6}$; $b_{51} = 0.0278$; $b_{52} = 4.13 \times 10^{-4}$.

Insulin Subsystem

The insulin subsystem is described by a five-compartment model, involving pancreatic insulin storage, liver and portal plasma insulin, plasma insulin and insulin in the interstitial fluid.

$$\dot{u}_{11}(t) = - (m_{01} + m_{21} + m_{31}) u_{11} + m_{12}u_{12} + m_{13}u_{13} + I_u(t), \quad u_{11}(0) = u_{110}$$
$$\dot{u}_{12}(t) = - (m_{02} + m_{12}) u_{12} + m_{21}u_{11} + k_{02}(x_1)u_{2p}, \quad u_{12}(0) = u_{120}$$
$$\dot{u}_{13}(t) = -m_{13}u_{13} + m_{31}u_{11}, \quad u_{13}(0) = u_{130}$$

where u_{11} = the quantity of insulin in plasma [μU]; u_{12} = the quantity of insulin in the liver [μU]; u_{13} = the quantity of insulin in the interstitial fluid [μU]; $I_u(t)$ = the insulin test input. The constants are $m_{01} = 0.125$, $m_{02} = 0.185$, $m_{12} = 0.209$, $m_{13} = 0.02$, $m_{21} = 0.268$, $m_{31} = 0.042$ (values for a normal state). The term $k_{02}(x_1)u_{2p} = F_6(u_{2p}, x_1)$ represents the insulin secretion rate.

- Insulin synthesis

$$W(x_1) = 0.5a_w \{1 + \tanh [b_w (e_x + c_w)]\}$$

Baseline parameter values for a normal state: $a_w = 0.287$; $b_w = 0.0151$; $c_w = -92.3$.

- Insulin secretion

$$F_6(u_{2p}, x_1) = 0.5a_6 \left\{1 + \tanh\left[b_6\left(e_x + c_6\right)\right]\right\} u_{2p}$$

Baseline parameter values for a normal state: a_6=1.3; b_6=0.0923; c_6=−19.68.

Glucagon Subsystem

The glucagon subsystem is described by a one-compartment model:

$$\dot{u}_2(t) = -h_{02}u_2 + F_7(x_1, u_{13}), \qquad u_2(0) = u_{20}$$

where u_2 = the quantity of glucagon in the plasma and interstitial fluid [ng]; F_7 = the endogenous release of glucagon, dependent on BG and interstitial fluid insulin; h_{02}=0.086.

- Glucagon secretion

$$F_7(x_1, u_{13}) = a_{71} H_7(u_{13}) M_7(x_1)$$

  where

$$H_7(u_{13}) = 0.5 \left\{1 - \tanh\left[b_{71}\left(e_{13} + c_{71}\right)\right]\right\}$$
$$M_7(x_1) = 0.5 \left\{1 - \tanh\left[b_{72}\left(e_x + c_{72}\right)\right]\right\}$$

Baseline parameter values for a normal state: a_{71}=2.35; b_{71}=6.86×10^{-3}; b_{72}=0.03; c_{71}=99.2; c_{72}=40.

W and $F_1 - F_7$ are nonlinear functions, m_{ij}, h_{ij}, and k_{ij} are constant rate parameters [min^{-1}] with the exception of k_{02} which is a function of x_1. See [202–204] for details on the identification of individual parameters, and their nominal values. The nominal values are shown in Appendix B. See also [215, 216] for extensions based on Cobelli et al's model.

4.4.6 Candas and Radziuk

Candas & Radziuk [199] used this non-linear model to aid in the design of an adaptive controller for the maintenance of basal glycaemia during euglycaemic hyperinsulinemic clamps:

$$\frac{dG(t)}{dt} = -\left[k_o + k(t)\right] G(t) + RG(t)$$
$$\frac{dk(t)}{dt} = -a_1 \cdot k(t) + a_2 \cdot i(t)$$
$$\frac{di(t)}{dt} = -a_3 \cdot i(t) + a_4 \cdot k(t) + a_6 \cdot i_3(t) + RI(t)$$

$$\frac{di_3(t)}{dt} = -a_6 \cdot i_3(t) + a_5 \cdot i(t)$$

where $G(t)$ = plasma glucose concentration [mg/dl]; k_o = insulin-independent fractional removal rate of glucose [min^{-1}]; $k(t)$ = insulin-dependent fractional removal rate of glucose [min^{-1}]; $i(t)$ [μU] = insulin mass in the central compartment; $i_3(t)$= insulin mass in a peripheral compartment non-active in glucose removal [μU] ; a_1–a_6 = fractional transfer rates of the three-compartment model of insulin kinetics (a_1, a_3, a_5, a_6 have units in min^{-1}, and a_2 has unit of min$^2/\mu$U, and $a_4$is in μU/(min^2)). The values for $\{a_i\}$ are 0.394, 0.142, 0.251, 0.394, 3.15×10^{-8}, and 2.8×10^3, while k_{sc} was assumed to be 0.03 min^{-1}.

RG(t) accounts for the rate of appearance of glucose in the systemic circulation from both the endogenous and exogenous sources (in [mg/ml · min]), and RI(t) represents the entry of both the endogenous and exogenous insulin into the systemic circulation (in [μU/min]).

4.4.7 Hovorka's Model

Hovorka et al [211] has originally developed a glucoregulatory model to represent glucose kinetics during basal conditions and during the intravenous glucose tolerance test in healthy subjects [210, 221]. The model was extended to represent the relationship between subcutaneous insulin infusion as input and intravenous glucose concentration as output, as part of ADICOL project (an European Union initiative).

The extended model consists of a glucose subsystem (glucose absorption, distribution/transport and disposal), an insulin subsystem (insulin absorption, distribution and disposal) and an insulin action subsystem (insulin action on glucose transport, disposal and endogenous production).

Hovorka's model, as according to [211] and [210] is presented below:

Glucose Subsystem

The core model is a two-compartment representation of glucose kinetics

$$\frac{dQ_1(t)}{dt} = -\left[\frac{F_{01}^c}{V_G G(t)} + x_1(t)\right] Q_1(t) + k_{12}Q_2(t) - F_R + U_G(t) + \text{EGP}_0\left[1 - x_3(t)\right]$$

$$\frac{dQ_2(t)}{dt} = x_1(t)Q_1(t) - \left[k_{12} + x_2(t)\right] Q_2(t); \qquad y(t) = G(t) = \frac{Q_1(t)}{V_G}$$

where Q_1 = masses of glucose in the accessible compartment (i.e. where measurements are made) [mmol]; Q_2 = masses of glucose in the non-accessible compartment [mmol]; k_{12} = transfer rate constant from the non-accessible to the accessible compartment [min^{-1}]; V_G = distribution volume of the accessible compartment [Litre]; y and G is the (measurable) glucose concentration [mmol/l], and EGP$_0$ = endogenous glucose production extrapolated to the zero insulin concentration [mmol/min].

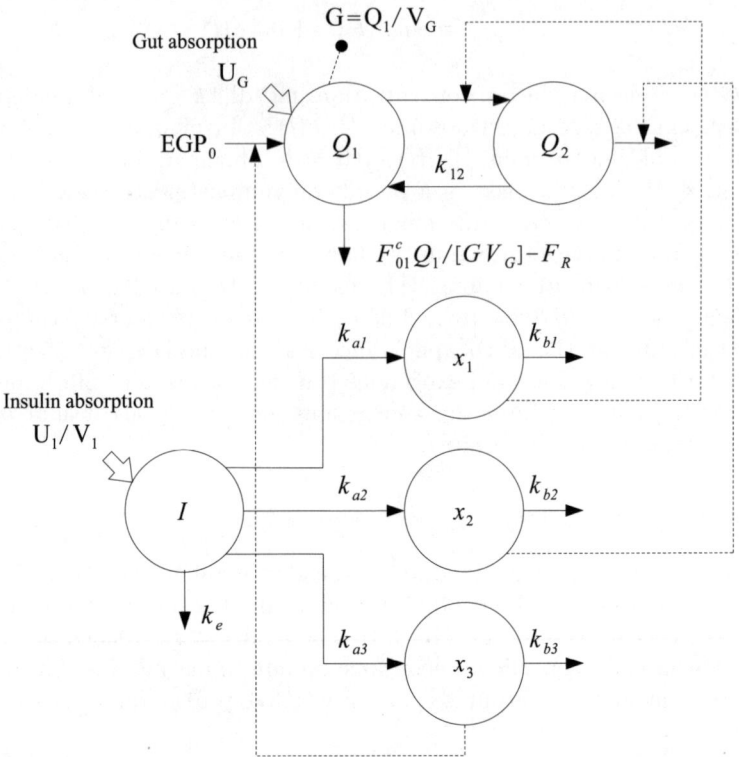

Fig. 4.4. Hovorka's model (Adapted from Hovorka et al [211])

F_{01}^c is the total insulin-independent glucose flux, corrected for the ambient glucose concentration [mmol/min]:

$$F_{01}^c = \begin{cases} F_{01} & \text{if } G \geq 4.5 \,\text{mmol/l} \\ F_{01}G/4.5 & \text{otherwise} \end{cases}$$

F_R is the renal glucose clearance above the glucose threshold of 9 mmol/l

$$F_R = \begin{cases} 0.003(G-9)V_G & \text{if } G \geq 9 \,\text{mmol/l} \\ 0 & \text{otherwise} \end{cases}$$

The gut absorption rate U_G is represented by a two-compartment chain with identical transfer rates of $1/t_{max,G}$

$$U_G(t) = \frac{D_G A_G \cdot t \exp(-t/t_{max,G})}{t_{max,G}^2}$$

where $t_{max,G}$ = the time-maximum appearance rate of glucose in the accessible glucose compartment; D_G = amount of carbohydrate digested; A_G = carbohydrate bio-availability; U_G in [mmol/min].

Insulin Subsystem

Insulin absorption is described as

$$\frac{dS_1(t)}{dt} = u(t) - \frac{S_1(t)}{t_{max,I}}$$

$$\frac{dS_2(t)}{dt} = \frac{S_1(t)}{t_{max,I}} - \frac{S_2(t)}{t_{max,I}}$$

where S_1, S_2 are a two-compartment chain representing absorption of subcutaneously administered short-acting insulin (e.g. Lispro); $u(t)$ = administration of insulin (bolus and infusion); $t_{max,I}$ = time-to-maximum insulin absorption. The insulin absorption rate (i.e. appearance of insulin in plasma) can be obtained as

$$U_1 = \frac{S_2(t)}{t_{max,I}}$$

The plasma insulin concentration $I(t)$ [mU/l] is described by

$$\frac{dI(t)}{dt} = \frac{U_I(t)}{V_I} - k_e I(t)$$

where k_e = fractional elimination rate [min^{-1}], and V_I is the distribution volume [L].

Insulin Action Subsystem

The three insulin actions on glucose kinetics are represented by

$$\frac{dx_1}{dt} = -k_{a1}x_1(t) + k_{b1}I(t), \qquad x_1(0) = \frac{k_{b1}}{k_{a1}}I_b$$

$$\frac{dx_2}{dt} = -k_{a2}x_2(t) + k_{b2}I(t), \qquad x_2(0) = \frac{k_{b2}}{k_{a2}}I_b$$

$$\frac{dx_3}{dt} = -k_{a3}x_3(t) + k_{b3}I(t), \qquad x_3(0) = \frac{k_{b3}}{k_{a3}}I_b$$

where x_1, x_2, x_3 = (remote) effects of insulin on glucose distribution/ transport, glucose disposal and endogenous glucose production, respectively [min^{-1}]; k_{a1}, k_{a2}, k_{a3} = deactivation rate constants [min^{-1}]; k_{b1}, k_{b2}, k_{b3} = activation rate constants [min^{-2} per mU/l]; I_b = basal plasma insulin [mU/l].

In conjunction with this model, a non-linear Model Predictive Controller with on-line parameter estimation has been developed (see Section 4.5.4). The following parameterisation has been made for the controller:

- Insulin sensitivity of distribution/transport

$$S_{IT}^f = \frac{k_{b1}}{k_{a1}}$$

- Insulin sensitivity of disposal

$$S_{ID}^f = \frac{k_{b2}}{k_{a2}}$$

- Insulin sensitivity of EGP

$$S_{IE}^f = \frac{k_{b3}}{k_{a3}}$$

The model quantities were divided into model constants and model parameters, mainly to reduce the number of parameters to be calculated, and at the same time, to be able to represent a wider range of glucose excursion observed in diabetic patients [211]. Some values for these model quantities can be found in Appendix B.

4.4.8 Sorensen's Model

The model presented here is mainly adapted from [209], with some parameters from [207] and [208], which are originally based on Sorensen's work [217, 222]. Using the compartmental modelling technique, the patient model is represented in Fig 4.5. Individual compartment models were obtained by performing mass balances around tissues important to glucose or insulin dynamics. There are six compartments, i.e. brain, heart/lungs, gut (combined effects of stomach and intestine), liver, kidney, and periphery (combined effects of muscle and adipose tissue). Blood transported glucose or insulin to the various compartments. It was assumed that the glucose or insulin concentration in a compartment was in equilibrium with the blood leaving the given compartment.

The nomenclature and parameter values can be found in [209] and are repeated in Appendix B.

Mass Balance for Glucose

- Brain:

$$\frac{dG_{BV}}{dt} = \frac{Q_B^G}{V_{BV}^G}(G_H - G_{BV}) - \frac{V_{BI}}{V_{BV}^G T_B}(G_{BV} - G_{BI})$$

$$\frac{dG_{BI}}{dt} = \frac{V_{BI}}{V_{BI} T_B}(G_{BV} - G_{BI}) - \frac{\Gamma_{BGU}}{V_{BI}}$$

- Heart and lungs:

$$\frac{dG_H}{dt} = \frac{1}{V_H^G} \cdot \left(Q_B^G G_{BV} + Q_L^G G_L + Q_K^G G_K + Q_P^G G_{PV} - Q_H^G G_H - \Gamma_{RBCU} \right)$$

- Gut:

$$\frac{dG_G}{dt} = \frac{Q_G^G}{V_G^G} \cdot (G_H - G_G) + \frac{1}{V_G^G} \cdot (\Gamma_{meal} - \Gamma_{GGU})$$

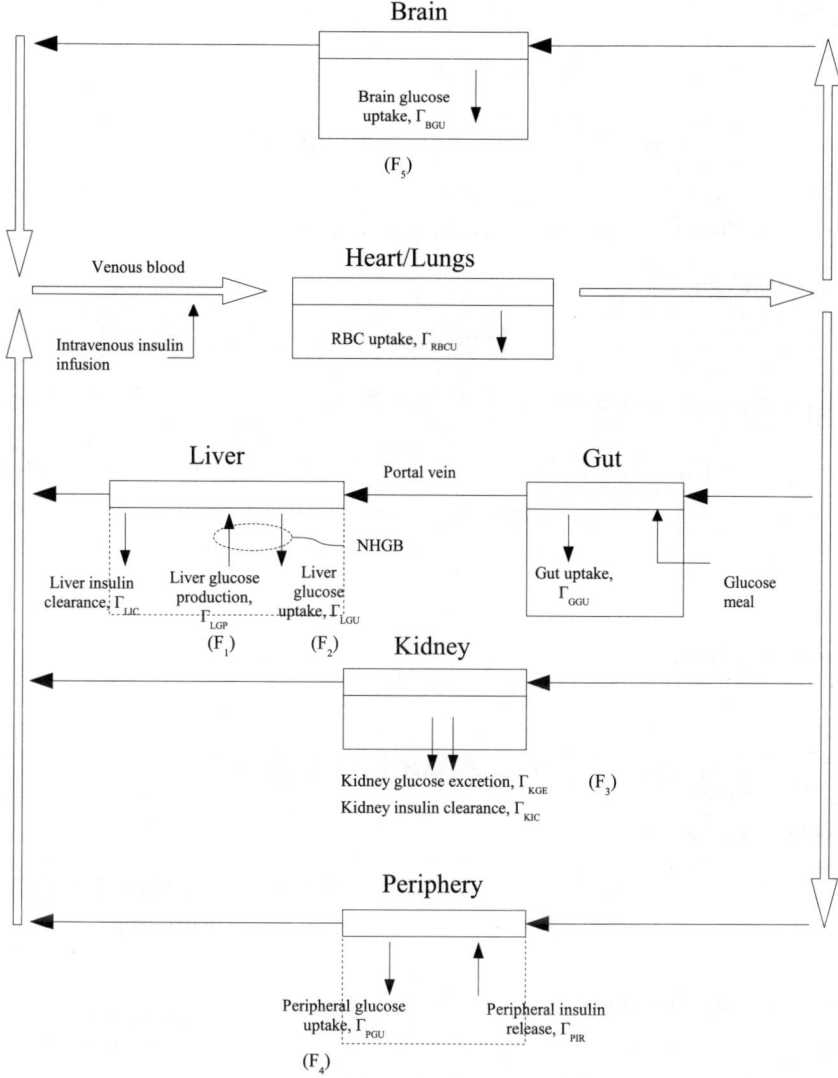

Fig. 4.5. Comprehensive glucose-insulin modelling (Adapted from [207, 208])

- Liver:

$$\frac{\mathrm{d}G_L}{\mathrm{d}t} = \frac{1}{V_L^G} \cdot \left(Q_A^G G_H + Q_G^G G_G - Q_L^G G_L + \Gamma_{HGP} - \Gamma_{HGU} \right)$$

- Kidney:

$$\frac{\mathrm{d}G_K}{\mathrm{d}t} = \frac{Q_K^G}{V_K^G} \cdot (G_H - G_K) - \frac{\Gamma_{KGE}}{V_K^G}$$

- Periphery:

$$\frac{\mathrm{d}G_{PV}}{\mathrm{d}t} = \frac{Q_P^G}{V_{PV}^G} \cdot (G_H - G_{PV}) - \frac{V_{PI}}{T_P^G V_{PV}^G} \cdot (G_{PV} - G_{PI})$$

$$\frac{\mathrm{d}G_{PI}}{\mathrm{d}t} = \frac{V_{PI}}{T_P^G V_{PI}} \cdot (G_{PV} - G_{PI}) - \frac{\Gamma_{PGU}}{V_{PI}}$$

Metabolic Sources and Sinks (Glucose Compartment)

- Peripheral glucose uptake:

$$\Gamma_{PGU} = \Gamma_{PGU}^B \cdot G_{PI}^N \left\{ 7.03 + 6.52 \tanh \left[0.388 \left(I_{PI}^N - 5.82 \right) \right] \right\}$$

- Hepatic glucose production:

$$\Gamma_{HGP} = \Gamma_{HGP}^B M_{HGP}^I \left\{ 2.7 \tanh \left(0.39 \chi^N \right) - f_2 \right\} \left\{ 1.42 - 1.41 \tanh \left[0.62 \left(G_L^N - 0.497 \right) \right] \right\}$$

$$\frac{\mathrm{d}}{\mathrm{d}t} (M_{HGP}^I) = \frac{1}{\tau_I} \left\{ 1.21 - 1.14 \tanh \left[1.66 \left(I_L^N - 0.89 \right) \right] - M_{HGP}^I \right\}$$

$$\frac{\mathrm{d}f_2}{\mathrm{d}t} = \frac{1}{\tau_\chi} \left(\frac{2.7 \tanh(0.39 \chi^N) - 1}{2} - f_2 \right)$$

- Hepatic glucose uptake:

$$\Gamma_{HGU} = \Gamma_{HGU}^B M_{HGU}^I \left\{ 5.66 - 5.66 \tanh \left[2.44 \left(G_L^N - 1.48 \right) \right] \right\}$$

$$\frac{\mathrm{d}}{\mathrm{d}t} (M_{HGP}^I) = \frac{1}{\tau_I} \left[2.0 \tanh \left(0.55 I_L^N \right) - M_{HGU}^I \right]$$

- Kidney glucose excretion:

$$\Gamma_{KGE} = \begin{cases} 71 + 71 \tanh \left[0.11 \left(G_K - 460 \right) \right], & 0 \leq G_K < 460\,\mathrm{mg/dl} \\ -330 + 0.872 G_K, & G_K \geq 460\,\mathrm{mg/dl} \end{cases}$$

Mass Balance for Insulin

- Brain:

$$\frac{\mathrm{d}I_B}{\mathrm{d}t} = \frac{Q_B^I}{V_B^I} (I_H - I_B)$$

- Heart and lungs:

$$\frac{\mathrm{d}I_H}{\mathrm{d}t} = \frac{1}{V_H^I} \cdot \left(Q_B^I I_B + Q_L^I I_L + Q_K^I I_K + Q_P^I I_{PV} - Q_H^I I_H + i(t) \right)$$

- Gut:

$$\frac{\mathrm{d}I_G}{\mathrm{d}t} = \frac{Q_G^I}{V_G^I} \cdot (I_H - I_G)$$

- Liver:

$$\frac{dI_L}{dt} = \frac{1}{V_L^I} \cdot \left(Q_A^I I_H + Q_G^I I_G - Q_L^I I_L + \Gamma_{PIR} - \Gamma_{LIC} \right)$$

- Kidney:

$$\frac{dI_K}{dt} = \frac{Q_K^I}{V_K^I} \cdot (I_H - I_K) - \frac{\Gamma_{KIC}}{V_K^I}$$

- Periphery:

$$\frac{dI_{PV}}{dt} = \frac{Q_P^I}{V_{PV}^I} \cdot (I_H - I_{PV}) - \frac{V_{PI}}{T_P^I V_{PV}^I} \cdot (I_{PV} - I_{PI})$$

$$\frac{dI_{PI}}{dt} = \frac{V_{PI}}{T_P^I V_{PI}^I} \cdot (I_{PV} - I_{PI}) - \frac{\Gamma_{PIC}}{V_{PI}}$$

Metabolic Sources and Sinks (Insulin Compartment)

- Liver insulin clearance:

$$\Gamma_{LIC} = F_{LIC} \left(Q_A^I I_H + Q_G^I I_G + \Gamma_{PIR} \right)$$

where $F_{LIC} = 0.40$

- Peripheral insulin clearance:

$$\Gamma_{PIR} = 0.0 \quad \text{(no pancreatic insulin release)}$$

- Kidney insulin clearance:

$$\Gamma_{KIC} = F_{KIC} \left(Q_K^I I_K \right)$$

where $F_{KIC} = 0.30$

- Peripheral insulin clearance:

$$\Gamma_{PIC} = \frac{I_{PI}}{\left[\left(\frac{1 - F_{PIC}}{F_{PIC}} \right) \left(\frac{1}{Q_P^I} \right) - \left(\frac{T_P^I}{V_{PI}} \right) \right]}$$

where $F_{PIC} = 0.15$

Glucagon Model

$$\frac{d\chi^N}{dt} = \frac{1}{V_\chi} \left(\Gamma_{M_\chi C} \cdot \Gamma_{P_\chi R}^N - \Gamma_{M_\chi C} \cdot \chi^N \right)$$

- Pancreatic glucagon release:

$$\Gamma_{P_\chi R}^N = \left\{ 2.93 - 2.10 \tanh \left[4.18 \left(G_H^N - 0.61 \right) \right] \right\} \cdot$$

$$\left\{ 1.31 - 0.61 \tanh \left[1.06 \left(I_H^N - 0.47 \right) \right] \right\}$$

where $\Gamma_{M_\chi C} = 0.91$ L/min, and $V_\chi = 9.94$ L.

4.4.9 Sub-models That Complements the Comprehensive Models

- Subcutaneously injected insulin:
  A first order model can be added to describe the transfer of the insulin mass from the subcutaneous (s.c.) depot to the systemic circulation [199]:

$$\frac{dM(t)}{dt} = -k_{sc} \cdot M(t) + \mathrm{RI}_{sc}(t)$$

$$\mathrm{IR}(t) = k_{sc} \cdot M(t)$$

where $M(t)$ = insulin mass in the s.c. depot $[\mu U]$; $\mathrm{RI}_{sc}(t)$ = insulin rate of appearance in the s.c. depot $[\mu U/\min]$; $k_{sc}(t)$ = fractional transfer rate from s.c. depot toward the systemic circulation $[\min^{-1}]$.

- Rate of glucose absorption via the gut wall:
  Lynch et al [223] cited the equations for the rate of glucose absorption from the gut wall [in mg/min]:

$$\mathrm{RG}_{abs} = K_{gabs}G_{gut}$$

where G_{gut} = the amount (mg) of glucose in the gut following ingestion of a meal and is defined by the following differential equation:

$$\frac{dG_{gut}}{dt} = \mathrm{RG}_{empty} - K_{gabs}G_{gut}$$

where RG_{empty} = the rate of gastric emptying which is described by a trapezoidal function, saturating at a certain maximal rate. This model was evaluated by Yates et al [224].

- Subcutaneous glucose infusion:
  A linear, first-order system in the frequency domain was introduced in [225] to be used in conjunction with Cobelli et al's model:

$$G_{sc}(s) = \frac{K_{sc}}{1 + t_{sc}s} \cdot \exp(-t_m s)$$

where K_{sc} = the gain; t_{sc} = time delay between blood glucose and s.c. glucose (5 min); t_m = time delay due to the tubing in an ex-vivo monitoring system (0–20 min).

- Subcutaneous glucose sensor:
  Rebrin et al [25] presented a model describing ISF glucose-sensor signal dynamics, characterized to allow for a gradient between plasma and interstitial glucose:

$$\frac{dC_2}{dt} = -\left(k_{02} + k_{12}\right) C_2 + k_{21}\frac{V_1}{V_2}C_1$$

where C_1 and C_2 are plasma and ISF glucose kinetics respectively, and V_1 and V_2 are plasma and ISF volume respectively. By substituting sensor current

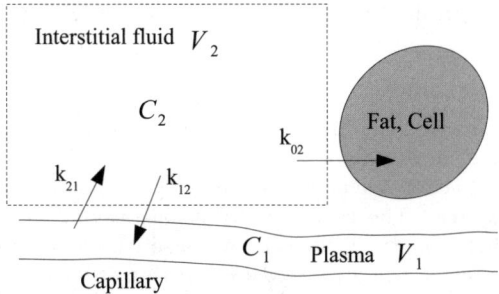

Fig. 4.6. ISF glucose-sensor signal dynamics (Adapted from [25])

$s = \alpha C_2$, the sensor signal can be described by

$$\frac{\mathrm{d}s}{\mathrm{d}t} = -p_2 s + p_3 C_1$$

where $p_2 = k_{02} + k_{12}$ and $p_3 = k_{21} V_1 / V_2 \alpha$ which can be identified from plasma glucose and sensor current data.

Quotes

Quote from [168], "It is clear that metabolic processes exhibit complex dynamics and involve many levels of control action ranging from autoregulatory effects occurring within the cell to hormonal control of metabolic systems at the global level. The approach that should be adopted in modelling metabolic system is conditioned, along with purpose and theory, the nature of the data available for system identification and model validation."

4.5 Mathematics of the Model-Based Control Algorithms

As previously stated, the availability of glucose-insulin model allows glucose-control problem to be treated as mathematical problems, and solutions (control algorithms) can be derived using various mathematical techniques. For linear glucose-insulin model, analytical results are obtainable in most cases. For non-linear models, the models are either linearised to reduce the number of parameters to be determined (and take advantage of the available mathematical techniques for linear models), or the control problem are solved numerically.

4.5.1 Pole Placement

This is a standard technique used in control systems design. It can be shown that if the system considered is completely state controllable, then the poles of the closed-loop system may be placed at any desired locations by means of state

Table 4.6. Definitions from Ricker [226]

- Plant:

  This represents the real "patient". The states of a plant are never fully measureable. One would not know the "parameters" or even the "order" of the plant, since it is generally a time-varying, non-linear, distributed-parameter system. The values of the signals leaving the controller are known, and the plant outputs can be measured, but nothing else in the plant is accessible. *The plant model is not part of the control system.* Its is only used to represent the plant in simulations and analytical work.

- Internal model:

  The internal model provides a prediction of future plant outputs as a function of contemplated adjustments in the manipulated variables and estimated disturbances, The controller chooses the values of control-inputs to send to the plant such that the predicted plant outputs are optimal according to a specified criterion. *The internal model is part of the control system*, and its states are all known exactly. Furthermore, the structure and parameters of the internal model are known.

feedback through an appropriate state feedback gain matrix [213]. Consider a control system

$$\dot{\mathbf{x}}(t) = \mathbf{A}\mathbf{x} + \mathbf{B}u$$

where $\mathbf{x}=$ state vector (n-vector); $u=$ scalar control signal; $\mathbf{A} = $ n $\times$ n constant matrix; $\mathbf{B} = $ n $\times$ 1 constant matrix. If we choose $u = -\mathbf{K}\mathbf{x}$, then we have

$$\dot{\mathbf{x}}(t) = (\mathbf{A} - \mathbf{B}\mathbf{K})\,\mathbf{x}(t)$$

If matrix $\mathbf{K}$ is chosen properly, the matrix $\mathbf{A} - \mathbf{B}\mathbf{K}$ can be made asymptotically stable, and for all $x(0) \neq 0$, it is possible to make $\mathbf{x}(t)$ approach 0 as t approaches infinity.

4.5.2 Optimal Control

In general, the optimal control problem attempts to find a feedback controller $\mathbf{u}(t)$ such that a linear system of the form (4.1) is satisfied, and a quadratic performance criterion

$$J(\mathbf{u}) = \int_0^\infty (\mathbf{x}(t) - x_d)^2 \, dt$$

is minimised. Minimising the criterion is equivalent to making $\mathbf{x}(t)$ as close to x_d as possible. x_d is a fixed constant from biological considerations.

Mathematically, the optimal control problem can be stated as

$$\min_{\mathbf{u}(t)} \int_0^\infty (\mathbf{x}(t) - x_d)^2 \, dt$$

The application of optimal control theory in glucose control was reported in several papers (e.g. [227, 228]). In these manuscript, a quadratic performance criterion of the form

$$J(u) = \int_0^\infty \left[x_1^2(t) + \rho u^2(t) \right] dt$$

was generally used on linear internal models (e.g. Ackerman's), where:

- $x_1(t)$ = glucose level above a basal value (i.e. $x(t) - x_d$)
- $\rho > 0$ = a scalar weighting factor included to adjust the sensitivity of the control to be achieved. Large values of ρ penalizes the use of large values of $u(t)$.
- both the glucose level and insulin usage are included in the performance index to be minimized.

A minimizing controller can generally be found for the linear system above (see Table 4.7):

Implementation in Literature

- Swan [228], and Fisher & Teo [227]: Both consider Ackerman's linear model as internal model,

$$\dot{x}_1(t) = -m_1 x_1(t) - m_2 x_2(t) + p(t)$$
$$\dot{x}_2(t) = m_4 x_1(t) - m_3 x_2(t) + u(t) \tag{4.9}$$

where $x_1 = G$ and $x_2 = H$.

However, Fisher & Teo lumped the two equations into one single equation

$$\ddot{g}_1 + (m_1 + m_3)\dot{g}_1 + m_1 m_3 x_1 = -m_2 u(t)$$

with initial condition $g_1(0) = x_{10}$ and $\dot{g}_1(0) = B - m_1 x_{10}$ and B = meal size [mg/dl per min]. The control problem is then solved as per usual method.

One observation from the solution to the glucose-control problem is that it is possible for $u(t)$ to become negative (i.e. the infusion of "negative" insulin). If we seek to infuse insulin only (and not be supplemented by glucose infusion to counter the previously given insulin), then the solution to the linear quadratic optimal control problem would not be suitable. Fisher & Teo suggested a "sub-optimal" control regime, where an extra constraint $u(t) \geq 0, \quad \forall t \geq 0$ to address this issue. Alternatively, the performance criterion can be re-formulated

$$J(u) = \int_0^\infty x_1^2(t) \, dt$$

subject to $0 \leq u(t) \leq u_{max}, \quad \forall t \geq 0$ (which could not be solved analytically [227]).

Table 4.7. Solving classical Optimal control problem

Solving Classical Optimal Control Problem

Given the system equation $\dot{\mathbf{x}} = \mathbf{A}\mathbf{x} + \mathbf{B}\mathbf{u}$ and performance index

$$J = \int_0^\infty [\mathbf{x}^*\mathbf{N}\mathbf{x} + \mathbf{u}^*\mathbf{R}\mathbf{u}]\,dt$$

where $\mathbf{N}$, $\mathbf{R}$ are both positive-definite Hermitian or real symmetric matrix. Let $\mathbf{u}(t) = -\mathbf{K}\mathbf{x}(t)$ where matrix $\mathbf{K}$ contains the optimal control vector, then $\dot{\mathbf{x}} = \mathbf{A}\mathbf{x} + \mathbf{B}\mathbf{u} = (\mathbf{A} - \mathbf{B}\mathbf{K})\,\mathbf{x}$ and

$$J = \int_0^\infty [\mathbf{x}^*\mathbf{Q}\mathbf{x} + \mathbf{u}^*\mathbf{R}\mathbf{u}]\,dt = \int_0^\infty \mathbf{x}^*\left[\mathbf{Q} + \mathbf{K}^*\mathbf{R}\mathbf{K}\right]\mathbf{x}\,dt$$

Choose a possible Liapunov function $V(\mathbf{x}) = \mathbf{x}^*\mathbf{P}\mathbf{x}$ where $\mathbf{P}$ is a positive definite Hermitian matrix. Since $V(\mathbf{x})$ is chosen to be a positive definite, then $\dot{V}(\mathbf{x})$ would be negative definite (for stability), $\dot{V}(\mathbf{x}) = -\mathbf{x}^*\mathbf{N}\mathbf{x}$. Note that:

$$\begin{aligned}
\dot{V}(\mathbf{x}) &= \dot{\mathbf{x}}^*\mathbf{P}\mathbf{x} + \mathbf{x}^*\mathbf{P}\dot{\mathbf{x}} \\
&= ((\mathbf{A} - \mathbf{B}\mathbf{K})\,\mathbf{x})^*\,\mathbf{P}\mathbf{x} + \mathbf{x}^*\mathbf{P}\,(\mathbf{A} - \mathbf{B}\mathbf{K})\,\mathbf{x} \\
&= \mathbf{x}^*\left[(\mathbf{A} - \mathbf{B}\mathbf{K})^*\,\mathbf{P} + \mathbf{P}\,(\mathbf{A} - \mathbf{B}\mathbf{K})\right]\mathbf{x} \\
&= -\mathbf{x}^*\mathbf{N}\mathbf{x}
\end{aligned}$$

Applying this to the performance index J, we obtained

$$(\mathbf{A} - \mathbf{B}\mathbf{K})^*\,\mathbf{P} + \mathbf{P}\,(\mathbf{A} - \mathbf{B}\mathbf{K}) = -\left[\mathbf{Q} + \mathbf{K}^*\mathbf{R}\mathbf{K}\right] \qquad (4.8)$$

Since $\mathbf{R}$ is assumed to be a positive-definite Hermitian or real symmetric matrix, we can write $\mathbf{R} = \mathbf{T}^*\mathbf{T}$ and re-write equation 4.8 to be

$$\mathbf{A}^*\mathbf{P} + \mathbf{P}\mathbf{A} - \mathbf{B}^*\mathbf{K}^*\mathbf{P} - \mathbf{P}\mathbf{B}\mathbf{K} + \left[\mathbf{Q} + \mathbf{K}^*\mathbf{T}^*\mathbf{T}\mathbf{K}\right] = 0$$

$$\mathbf{A}^*\mathbf{P} + \mathbf{P}\mathbf{A} + \left[\mathbf{T}\mathbf{K} - \frac{1}{\mathbf{T}^*}\mathbf{B}^*\mathbf{P}\right]^*\left[\mathbf{T}\mathbf{K} - \frac{1}{\mathbf{T}^*}\mathbf{B}^*\mathbf{P}\right] - \mathbf{P}\mathbf{B}\mathbf{R}^{-1}\mathbf{B}^*\mathbf{P} + \mathbf{Q} = 0$$

For the equation above to be the minimum requires $\mathbf{T}\mathbf{K} - \frac{1}{\mathbf{T}^*}\mathbf{B}^*\mathbf{P} = 0$, or

$$\begin{aligned}
\mathbf{K} &= \frac{1}{\mathbf{T}}\frac{1}{\mathbf{T}^*}\mathbf{B}^*\mathbf{P} \\
&= \frac{1}{\mathbf{R}}\mathbf{B}^*\mathbf{P}
\end{aligned}$$

Thus, the optimal control law is $\mathbf{u}(t) = -\mathbf{K}\mathbf{x}(t) = -\mathbf{R}^{-1}\mathbf{B}^*\mathbf{P}\mathbf{x}(t)$.
The optimal control problem can also be solved using a Hamiltonian (See Gopal [229]).

4.5.3 Adaptive Control

In adaptive control of glucose level, a glucose-insulin model is used as internal model, and the parameters of the model are first estimated using adaptive algorithms. Then other mathematical techniques (including optimal control) are applied onto this model that has the estimated parameters, for the eventual calculation of the insulin infusion rate. Fig 4.7 shows the general arrangement.

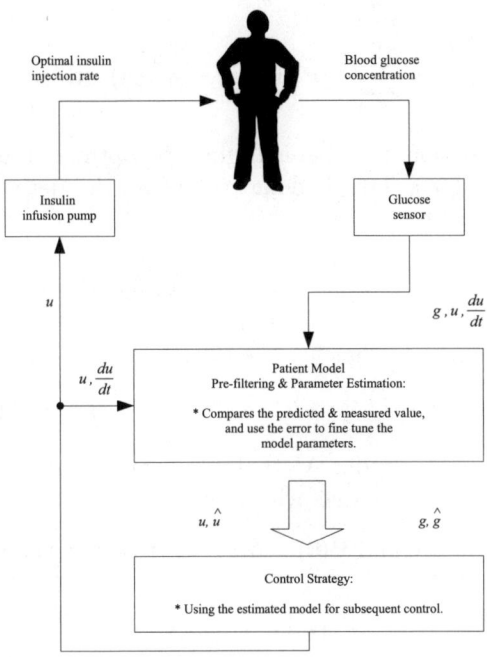

Fig. 4.7. Adaptive control general arrangement

The adaptive method usually considered in the literature is the (recursive) least square estimation algorithm.

Least Square Estimation Algorithm (from [230])

Consider an input samples from a time series
$\mathbf{x}(k) = [x(k), x(k-1), \cdots, x(k-p)]^T$, and the coefficients of a time-varying Finite Impulse Response filter $\mathbf{w}(n) = [w_0(n), w_1(n), \cdots, w_p(n)]^T$. Together, they represent the output of a system as $y(k) = \mathbf{w}^T(n) \cdot \mathbf{x}(k)$ at time k.

If $d(k)$ is the desired output at time k, then the error signals is

$$e(n, k) = d(k) - y(k) = d(k) - \mathbf{w}^T(n) \cdot \mathbf{x}(k)$$

By defining a forgetting factor $\beta(n,k)$ to reduce the influence of old data

$$0 < \beta(n,k) \leq 1, \quad k = 1, 2, \ldots, n$$

the error can be redefined as

$$\xi(n) = \xi\left(w_0(n), w_1(n), \cdots, w_p(n)\right)$$

$$= \sum_{k=0}^{n} \beta(n,k) \left|e(n,k)\right|^2$$

$$= \sum_{k=0}^{n} \beta(n,k) \left|d(k) - \mathbf{w}^T(n) \cdot \mathbf{x}(k)\right|^2$$

In least-square estimation, the goal is to find the optimal filter coefficients that minimises the error $\xi(n)$. This is done by setting the derivative of $\xi(n)$ with respect to $w_m^*(n)$ to zero:

$$\frac{\partial \xi(n)}{\partial w^*} = \sum_{k=0}^{n} \beta(n,k) e(n,k) \frac{\partial e^*(n,k)}{\partial \mathbf{w}^*} = 0$$

It is common to use $\beta(n,k)$ that has the form

$$\beta(n,k) = \lambda^{n-k}, \quad k = 1, 2, \ldots, n; \quad 0 < \lambda < 1$$

It can be shown (see Appendix A) that the optimal filter coefficients can be found using a recursion technique, which begins by

1. setting $\mathbf{w}^T(n-1) = 0$ and $\mathbf{P}(0) = \delta^{-1} I$ where δ is a small positive constant
2. computing

$$\mathbf{k}(n) = \frac{\lambda^{-1} \mathbf{P}(n-1) \mathbf{x}^*(n)}{1 + \lambda^{-1} \mathbf{x}^T(n) \mathbf{P}(n-1) \mathbf{x}^*}$$

$$\mathbf{w}(n) = \mathbf{w}(n-1) + \left(d(n) - \mathbf{x}^T(n) \mathbf{w}(n-1)\right) \mathbf{k}(n)$$

$$\mathbf{P}(n) = \lambda^{-1} \mathbf{P}(n-1) - \lambda^{-1} \mathbf{k}(n) \mathbf{x}^T(n) \mathbf{P}(n-1) \qquad (4.10)$$

3. Repeat until $\mathbf{w}(n)$ converges or continue indefinitely.

Implementation in Literature

• Pagurek et al [231] used a generalised least square algorithm to estimate parameters in Ackerman's linear model (i.e. m_1 to m_4) on-line, with signal pre-filtering. The recursive estimation of the parameters followed that of system 4.10, with the filter coefficients derived from a difference equation $y(n) + a_1 y(n-1) + a_2 y(n-2) = b_1 u(n-k-1) + b_2 u(n-k-2) + e(n)$ relating to Ackerman's model. The model, and a disturbance model (assumed to be samples of stationary independent Gaussian zero-mean sequences) are then used in a minimum variance control strategy.

- Kikuchi et al [232] used recursive least square estimation to estimate the parameters in the Ackerman's model (i.e. m_1 to m_4). The recursive estimation followed that of system 4.10 but with modification to include an extra term "$-(1 - m_1\mathrm{T})X_1(n - 2)$" in the term $d(n) - \mathbf{x}^T(n)\mathbf{w}(n - 1)$ of system 4.10 above. X_1 was defined to be the equivalent of glucose concentration x_1 in the Ackerman's equations. The filter coefficients were derived from a difference equation relating to Ackerman's model. A state observer was then used to estimate the insulin state variable before calculating the insulin infusion rate using optimal control strategy. The state observer is used because only the glucose (first state variable) and not the insulin (the second variable) in Ackerman's model is directly measured in real-time. Due to the delay in BG measurement from the glucose analyser, a state-predictor was included to deduce the present time BG (based on the BG taken some minutes ago).

- Sarti et al [233] devised a self-tuning adaptive controller based on a discrete-time linear glucose-insulin model P described by:

$$G_k = P\left(G_{k-1}, \ldots, G_{k-m},\ \mathrm{ID}_{k-1}, \ldots, \mathrm{ID}_{k-n}, \Theta\right)$$

where G_k = BG level at time k; ID_k = insulin dose at time k; Θ = a set of unknown parameters; and m, n = known time delays. The parameter set is recursively estimated at each time k based on G_k measurements to produce G_{k+1} (a one-step-ahead prediction). An optimal insulin dose, ID_{k+1} is calculated by minimizing a cost function J, which in the case of a minimum variance controller, is

$$J_k = (G_{k+1} - G_{desired})^2 - r \cdot \mathrm{ID}_{k+1}^2$$

with r being a weighting factor that penalises insulin dosage.

- Fischer et al [214] devised an adaptive controller with system identification. System identification was based upon a linear second-order difference equation model describing the actual state of the controlled plant of the glucose-insulin system (in terms of blood glucose concentration BG, and exogenous insulin dose ID) at discrete intervals k. A recursive least square regression analysis was applied to estimate the model parameters of this internal model. The performance index was

$$\Sigma\left(e^2(k) - 0.003\mathrm{ID}^2(k)\right)$$

where $e(k) = \mathrm{BG}(k) - \mathrm{BG}_{setpoint}$. The insulin dose is then given by

$$\mathrm{ID}(k + 1) = a_1^* \cdot e(k) + a_2^* \cdot e(k - 1) - a_3^* \cdot \mathrm{ID}(k - 1)$$

where a_i^* = estimated control constants.

- Candas & Radziuk [199] used an adaptive control based on a non-linear insulin-glucose model, and instead of infusing insulin only, glucose infusion is also considered. The approach counters the glucose removal (due to externally infused insulin) with an exogenous infusion of glucose. The externally infused insulin is kept at a constant rate, and blood glucose concentration is kept at a desired level by means of controlling an external infusion of glucose (glucose-clamp experiment). The process parameters on the non-linear model are recursively adjusted (via generalized-least-square method) after each new observation of the glucose concentration. Control is then achieved by using the nonlinear glucose kinetics model equation in an inverse fashion to calculate the glucose infusion required to maintain a preset profile for glucose.

4.5.4 Model Predictive Control

The MPC control law can be understood by referring to Fig 4.8. For any assumed set of present and future control moves $\Delta u(k)$, $\Delta u(k+1)$, ..., $\Delta u(k+m-1)$ the future behaviour of the process outputs $y(k+1|k)$, $y(k+2|k)$, ..., $y(k+p|k)$ can

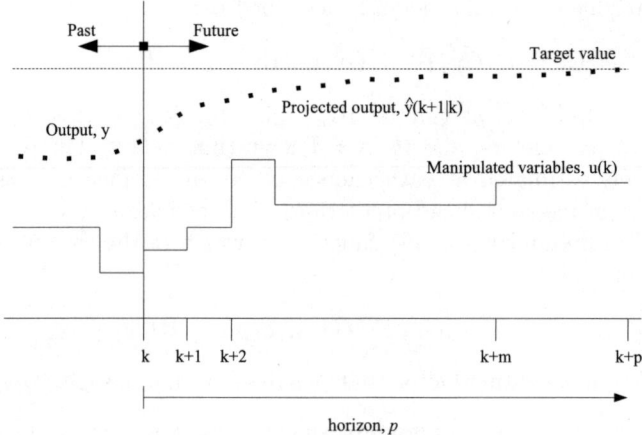

Fig. 4.8. Model Predictive control illustration (Adapted from [234])

be predicted over a horizon p. The m present and future control moves ($m \leq p$) are calculated to minimize a quadratic objective of the form

$$\min_{\Delta u(k)...\Delta u(k+m-1)} \sum_{n=1}^{p} \| \Gamma^y \left[y(k+n|k) - r(k+n) \right] \|^2 + \sum_{n=1}^{m} \| \Gamma^u \left[\Delta u(k+n-1) \right] \|^2$$

where $y(k+n|k)$ = predicted value of y at time $k+n$ based on information available at time k; $\Delta u(k+n) = u(k+n) - u(k+n-1)$; Γ^y and Γ^u = weighting matrices to penalise certain components of y or u at certain future time intervals; $r(k+n)$ = vector of future reference values (setpoints).

Although m control moves are calculated, only the first one (i.e. $\Delta u(k)$) is implemented. At the next sampling interval, new values of the measured output are obtained, the control horizon is shifted forward by one step, and the same calculations are repeated. The resulting control law is referred to as "moving horizon" or "receding horizon". The predicted process outputs $y(k+1|k)$, . . ., $y(k+p|k)$ depend on

- the current measurement $\hat{y}(k)$, and
- the assumptions made about the unmeasured disturbances and measurement noise affecting the outputs (see [234]).

In MPC, the goal is to find the best value for the manipulated variable $u(t)$. More details about MPC and the inclusion of disturbance into the calculation can be found in [235] and [234].

Implementation in Literature

- Parker et al [207] used linear Model Predictive Control with state estimation and Kalman filtering. The internal model is

$$x(k+1) = \Phi x(k) + \Gamma u(k)$$
$$y(k) = Cx(k)$$

The state $\hat{x}$ and output $\hat{y}$ of the plant can be estimated using

$$\hat{x}(k+1) = \Phi \hat{x}(k) + \Gamma u(k) + \kappa[y(k) - C\hat{x}(k)]$$
$$\hat{y}(k) = C\hat{x}(k)$$

where the Kalman filter κ is calculated iteratively off-line:

$$\kappa(i) = \Phi P(i)C^T \left(\frac{1}{CP(i)C^T + R_2} \right)$$
$$P(i+1) = \Phi P(i)\Phi^T + R_1 - \kappa(i)CP(i)\Phi^T$$

Tuning of the Kalman filter is accomplished using the matrices R_1, R_2 and P(0).
The internal model is obtained from the continuous nonlinear diabetic patient model by first linearising the nonlinear model analytically to produce a linear continuous-time model. The linear model is then converted to a discrete-time minimum-phase model for use in simulation, yielding Φ, Γ and C.

- Lynch et al [223] used a similar concept except that Bergman's model is discretized (instead of linearising the comprehensive model), and that subcutaneous glucose measurement is used (rather than the arterial glucose measurement in Parker's case).

- Trajanoski et al [225] devised a "Neural Predictive Controller", which is based on off-line identification of the glucoregulatory system using neural network, amalgamated with a nonlinear model predictive controller (nMPC) design using multiple step-ahead predictions of the previously identified model. The nMPC uses moving horizon approach with a whole set of predictions (instead of the classical approach where one prediction is used). The performance index was:

$$\min_{u(t),u(t+1)...u(t+m-1)} J = \Sigma_{j=N_1}^{p} \left(e_j^T \Gamma_e e_j \right) + \Sigma_{j=0}^{m-1} \left[\Delta u^T(t+j) \cdot \Gamma_u \cdot \Delta u(t+j) \right]$$

where e_j = deviation of the predicted output $\hat{y}(t+j)$ from the setpoints $r(t+j)$; $\Delta u = u(t) - u(t-1)$ is the difference in the control moves; m = control horizon; p = prediction horizon; N_1 = minimum costing horizon; Γ_e, Γ_u = prediction and control weighting respectively. Bellazi et al [236] explained this approach from a different perspective.

- Hovorka et al [211] have developed a non-linear model predictive controller with on-line parameter estimation using Bayesian approach. The model parameters S_{IT}^f, S_{ID}^f, S_{IE}^f, F_{01} and EGP_0 (see Section 4.4.7) are tuned using the objective function[2]

$$\arg \min_{-2.5 \leq p_1 \cdots p_6 \leq 2.5} \left\{ \sum_{i=1}^{N_w} w_{t-i} \left[\hat{y}\left(t-i|p_1 \cdots p_6\right) - y\left(t-i\right) \right]^2 + \sum_{k=1}^{6} p_k^2 \right\}$$

where w_i = weight reciprocal to the square of the measurement error (coefficient of variation of 2–9% depending on the source of glucose sample); $\hat{y}\left(t-i|p_1 \cdots p_6\right)$ = glucose concentration predicted by the model at time t given the standardised parameters $p_1 \cdots p_6$; N_w = length of the learning window; $y\left(t-i\right)$ = previously measured glucose concentration.

The parameters $p_1 \cdots p_6$ are independent normal distributions with a zero mean and a unit standard deviation (i.e. $p_1 \sim N(0,1)$, $i = 1, \ldots, 5$), and they are related to the glucoregulatory model by

$$\ln\left(S_{IT}^f\right) = a_{11}p_1 + b_1$$

$$\ln\left(S_{ID}^f\right) = a_{12}p_1 + a_{22}p_2 + b_2$$

$$\ln\left(S_{IE}^f\right) = a_{13}p_1 + a_{23}p_2 + a_{33}p_3 + b_3$$

$$\ln\left(F_{01}\right) = a_{14}p_1 + a_{24}p_2 + a_{34}p_3 + a_{44}p_4 + b_4$$

$$\ln\left(EGP_0\right) = a_{15}p_1 + a_{25}p_2 + a_{35}p_3 + a_{45}p_4 + a_{55}p_5 + b_5$$

with the coefficients a_{ij} and b_i derived from Random Variable Transformation technique. The insulin infusion rates are calculated by minimising an objective function over a prediction horizon N:

$$\arg \min_{0 \leq u(t+1) \cdots u(t+N) \leq 4} \left\{ \sum_{i=1}^{N} \left[\hat{y}\left(t+i|t\right) - y\left(t+i\right) \right]^2 \right.$$

$$\left. + \frac{1}{k_{agr}} \sum_{i=1}^{N} \left[u\left(t+i\right) - u\left(t+i-1\right) \right]^2 \right\}$$

where $\hat{y}\left(t+i|t\right)$ = predicted glucose; $y\left(t+i\right)$ = target (or desired) trajectory of glucose associated with high and low glucose concentration; k_{agr} = an "aggressive" constant balancing the contribution of the two summation terms above. An example of the target trajectory that the system would follow would be a BG curve that linearly decreases at pre-defined rates per

---

[2] Notation: $\arg \min_x f(x)$ is the value of x for which $f(x)$ has the minimum value.

hour (see Fig 4 in [211]). A sequence of insulin rates $u(t + 1), \ldots, u(t + N)$ are generated, but only the first one, $u(t + 1)$, is delivered to the patient. The insulin rate was limited to 4 U/hr due to the limitation of the infusion pump used (see [211]). Note that in clinical trials involving human subjects, blood glucose is measured intravenously (i.e. accurate BG values were obtained and fed into the controller) while insulin are given via a subcutaneous route.

4.5.5 H_∞ Control

H_∞ refers to the space of all bounded functions that are analytic and stable (i.e. poles are in the Right-hand-plane), and have proper transfer functions (i.e. degree of the denominator $\geq$ the degree of the numerator) (see [237]). Consider a stable single-input-single-output (SISO) linear system with transfer function $G(s)$, the H_∞ norm is defined as

$$\|G\|_\infty = \sup_\omega |G(j\omega)|$$

The interest is to optimise over the space of transfer function. Graphically, it is simply the peak in the Bode magnitude plot of the transfer function. H_∞ norm is a measure of a worst-case system gain.

Table 4.8. L_2-norm and L_∞-norm

Definition: L_2-norm & L_∞-Norm

For a scalar-valued signal $v(t)$ defined for $t \geq 0$ the L_2-norm is defined as the square root of the integral of $v(t)^2$:

$$\|v\|_2 = \left(\int_0^\infty v(t)^2 dt \right)^{1/2}$$

The L_2-norm is a special case of L_p-norm, defined as

$$\|v\|_p = \left(\int_0^\infty v(t)^p dt \right)^{1/p}, \quad p \geq 1$$

As $p \to \infty$, the L_p-norm tends to the so-called ∞-norm, or L_∞-norm, which is the least upper bound (or supremum) of the absolute value, i.e..

$$\|v\|_\infty = \sup_t |v(t)|$$

The *supremum* is used in place of *maximum* is because there is no guarantee that a maximum exists for $v(t)$ in general.

Table 4.9. Packed-Matrix Notation

Definition: Pack-Matrix Notation (See [238])

The transfer function of a system with state-space matrices [A, B, C, D] is given by

$$G(s) = C(sI - A)^{-1} + D$$

This transfer function in packed-matrix form is written as

$$G(s) = \left[\begin{array}{c|c} A & B \\ \hline C & D \end{array}\right]$$

Note that the packed-matrix form is the frequency domain transfer function expressed like a time-domain matrix, i.e.

$$G(s) = \left[\begin{array}{c|c} A & B \\ \hline C & D \end{array}\right]$$
$$= C(sI - A)^{-1} + D$$
$$= s - \text{domain transfer function}$$

Consider the standard configuration shown in Fig 4.9. Plant P has:

- Two inputs:
  1. w = the exogenous input, which includes the reference signal and disturbances;
  2. u = manipulated variables.

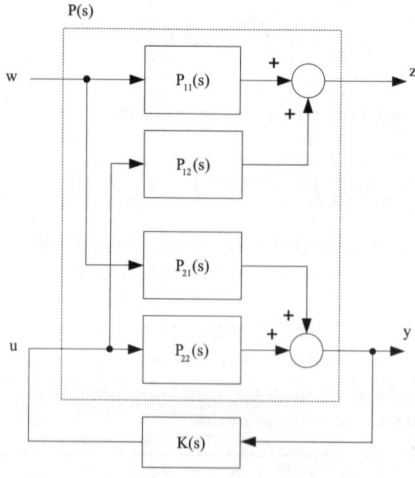

Fig. 4.9. H_∞ standard configuration (Adapted from [239, 240])

- Two outputs:
  1. z = error signals (which is what we wanted to minimise);
  2. y = measured variables (which is used in the controller K to calculate the manipulated variable u, so as to control the system).

P and K are matrices while the input & output signals are generally vectors. The system can be described by:

$$\begin{bmatrix} \text{tracking error} \\ \text{control output} \end{bmatrix} = P \begin{bmatrix} \text{reference signal} \\ \text{disturbances} \\ \text{control input} \end{bmatrix}$$

Let the SISO model P has a state-space system of the form

$$\frac{dx(t)}{dt} = Ax(t) + B_1 w(t) + B_2 u(t)$$
$$z(t) = C_1 x(t) + D_{11} w(t) + D_{12} u(t)$$
$$y(t) = C_2 x(t) + D_{21} w(t) + D_{22} u(t)$$

Some algebraic manipulation:

$$s\mathbf{x} = A\mathbf{x} + B_1\mathbf{w} + B_2\mathbf{u} \qquad\qquad \Rightarrow (sI - A)\mathbf{x} = B_1\mathbf{w} + B_2\mathbf{u}$$
$$\mathbf{z} = C_1\mathbf{x} + D_{11}\mathbf{w} + D_{12}\mathbf{u}$$
$$\mathbf{y} = C_2\mathbf{x} + D_{21}\mathbf{w} + D_{22}\mathbf{u}$$

and substituting $\mathbf{x} = \frac{1}{sI-A}B_1\mathbf{w} + \frac{1}{sI-A}B_2\mathbf{u}$ into the second and third equation, we obtain

$$\mathbf{z} = \left[C_1 \frac{1}{(sI - A)} B_1 + D_{11} \right] \mathbf{w} + \left[C_1 \frac{1}{(sI - A)} B_2 + D_{12} \right] \mathbf{u}$$
$$\mathbf{y} = \left[C_2 \frac{1}{(sI - A)} B_1 + D_{21} \right] \mathbf{w} + \left[C_2 \frac{1}{(sI - A)} B_2 + D_{22} \right] \mathbf{u}$$

or

$$\begin{bmatrix} \mathbf{z} \\ \mathbf{y} \end{bmatrix} = \begin{bmatrix} P_{11}(s) & P_{12}(s) \\ P_{21}(s) & P_{22}(s) \end{bmatrix} \begin{bmatrix} \mathbf{w} \\ \mathbf{u} \end{bmatrix} \tag{4.11}$$

$$= \left(\begin{bmatrix} D_{11} & D_{12} \\ D_{21} & D_{22} \end{bmatrix} + \begin{bmatrix} C_1 \\ C_2 \end{bmatrix} (sI - A)^{-1} \begin{bmatrix} B_1 & B_2 \end{bmatrix} \right) \begin{bmatrix} \mathbf{w} \\ \mathbf{u} \end{bmatrix}$$

$$= \begin{bmatrix} A & B_1 & B_2 \\ C_1 & D_{11} & D_{12} \\ C_2 & D_{21} & D_{22} \end{bmatrix} \begin{bmatrix} \mathbf{w} \\ \mathbf{u} \end{bmatrix} \qquad \text{by definition of packed matrix notation}$$

If a feedback controller of the form $u = K(s)y$ is introduced, the equations can be transformed as follows

$$\mathbf{z} = P_{11}\mathbf{w} + P_{12}\mathbf{u} \qquad \text{and} \qquad \mathbf{y} = P_{21}\mathbf{w} + P_{22}\mathbf{u}$$
$$= P_{11}\mathbf{w} + P_{12}K\mathbf{y} \qquad\qquad\qquad = P_{21}\mathbf{w} + P_{22}K\mathbf{y}$$
$$(I - P_{22}K)\mathbf{y} = P_{21}\mathbf{w}$$
$$\Rightarrow \quad \mathbf{y} = (I - P_{22}K)^{-1} P_{21}\mathbf{w}$$

We can express the transfer function of w-to-z as

$$\mathbf{z} = F(P, K) \cdot \mathbf{w}$$

where

$$F(P, K) = P_{11} + P_{21} K \frac{1}{(I - P_{22}K)} P_{21}$$

The expression for the closed-loop transfer function $F(P, K)$ is called Linear Fractional Transformation. The objective of H_∞ control design is to find a controller K that minimises the cost

$$J_\infty(K) = \|F(P, K)\|_\infty$$
$$= \sup_\omega \bar{\sigma}\left(F(P, K)(j\omega)\right)$$
$$= \sup_\omega \bar{\sigma}\left(\frac{\|z\|}{\|w\|}\right)$$

The direct minimisation of the cost $J_\infty(K)$ turns out to be difficult. It is much easier (and practical) to construct conditions which state whether there exists a stabilising controller such that the H_∞ norm is bounded

$$J_\infty(K) < \gamma$$

for a given $\gamma > 0$.

One interesting properties of H_∞ norm is that it gives the maximum factor by which a system magnifies the L_2 norm of any input [241],

$$J_\infty(K) = \sup_\omega \left\{ \frac{\|z\|_2}{\|w\|_2} : w \neq 0 \right\} < \gamma \qquad (4.12)$$

Let $L(w, u) = \|z\|_2^2 - \gamma^2 \|w\|_2^2 < 0, \quad \forall w \neq 0$, then by Parseval's theorem, the time-domain expression of the L_2 norm becomes

$$L(w, u) = \int_0^\infty \left[z(t)^T z(t) - \gamma^2 w(t)^T w(t) \right] dt < 0, \quad \forall w \neq 0$$

The problem of finding a controller $u = Ky$ can be stated in terms of a max-min problem

$$\sup_{w \neq 0} \left\{ \inf_{u=Ky} L(w, u) \right\} < 0$$

Put simply, w (player 1) will try to make the cost $L(u, w)$ as large as possible, while u (player 2) attempts to find the smallest $L(u, w) < 0$.

Certain assumptions must be satisfied for the H_∞ controller to exist [237,238]:

1. The pair (A, B_2) be controllable and (C_2, A) be detectable
2. Rank $D_{12}=$ dimension of u, and rank $D_{21}=$ dimension of y
3. Rank $\begin{bmatrix} A - j\omega I & B_2 \\ C_1 & D_{12} \end{bmatrix}$ = dimension of x + dimension of u
4. Rank $\begin{bmatrix} A - j\omega I & B_1 \\ C_2 & D_{21} \end{bmatrix}$ = dimension of x + dimension of y

A procedure called γ-iteration is used to find a stabilising controller: Start with a value for γ and reduce it until the problem failed to have a solution [237].

The design of linear H_∞ controller requires a linear system model. From literature survey, controller design used:

1. linear glucose-insulin models, or
2. linearised models, where
   - linearisation techniques were used to transform a non-linear model into a linear one, or
   - model reduction technique was used to generate a low-order process model for use in controller synthesis.

Implementation in Literature

- Kienitz et al [242] designed a full-observer/state-feedback-type H_∞ controller by first using a slightly extended linear Ackerman's model:

$$\frac{\mathrm{d}x}{\mathrm{d}t} = A_g x(t) + B_{1g} p(t) + B_{2g} u(t)$$
$$y(t) = C_g x(t)$$

where $p(t)$ = the absorption of glucose from blood; $u(t)$ = infused insulin; $x = \begin{bmatrix} G\ I\ C \end{bmatrix}^T$ is the state vector that has an extra component C; G, I and C represent the glucose concentration, insulin concentration and the aggregate of glucagon, cortisol and catecholamine in the blood, respectively. The system above may also be written as

$$y(s) = \left[C_g (sI - A)_g^{-1} B_{2g} \right] u(s) + \left[C_g (sI - A)^{-1} B_{1g} \right] p(s)$$
$$= G_1(s) u(s) + G_2(s) p(s)$$
$$= P_{22}(s) u(s) + P_{21}(s) w(s) \qquad \text{by analogy to equation 4.11}$$

We can see that $u(s)$ and $p(s)$ is analogous to u and w, just as G_1 and G_2 being analogous to P_{21} and P_{22} in equation 4.11. G_1 can be seen as the patient model with the insulin infusion, and G_2 being the meal disturbance model. Parameter uncertainties in A_g were then modelled by additive uncertainties

$$G_1(s) = G_{10}(s) + \triangle G_1(s)$$
$$G_2(s) = G_{20}(s) + \triangle G_2(s)$$

Table 4.10. Robustness and Uncertainty modelling

Robustness and Uncertainty Modelling

A robust control system is one that can withstand variations in real environment and operate properly in realistic situations [240]. Control systems are usually designed using simplified models of the system and environment, and thus may not work on the real plant in real environment. A robust controller must perform satisfactorily not just for one plant, but for a family of plants.

Model uncertainty is generally divided into structured uncertainty and unstructured uncertainty. Structured uncertainty assumes that the uncertainty is modelled and there are bounds for uncertain parameters in the system. Unstructured uncertainties assume less knowledge of the system, and assumed only that the frequency response of the system lies between two bounds. Two most common unstructured uncertainties are:

- Additive, where

$$\text{actual system} = \text{modelled system} + \text{model error}$$
$$\widetilde{G}(s) = G(s) + \Delta_a(s)$$

  The model error, or the *additive uncertainty*, is given by

$$\Delta_a(s) = \widetilde{G}(s) - G(s)$$

- multiplicative, where

$$\widetilde{G}(s) = G(s) + \Delta_m(s)G(s)$$
$$= [1 + \Delta_m(s)]\, G(s)$$

  The model error, or the *multiplicative uncertainty*, is given by

$$\Delta_m(s) = \frac{\widetilde{G}(s) - G(s)}{G(s)}$$

The multiplicative model is used more often because it represents the relative error in the model, rather than the absolute error as represented by the additive model.

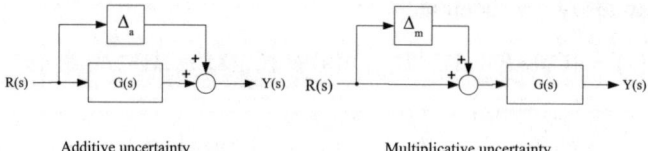

Additive uncertainty   Multiplicative uncertainty

For convenience, the characterisation of an uncertain plant is usually stated in an equivalent form obtained by defining a quantity "Δ" such that $\Delta_a = W_a\Delta$ for additive uncertainty, and $\Delta_m = W_m\Delta$ for multiplicative uncertainty [240]. The uncertainty block Δ is assumed to be uniformly bounded at all frequencies, and the varying amount of plant uncertainty at various frequencies is completely incorporated into the filter W_n. The filter W_n is commonly called "uncertainty weighting filter".

The design was refined by adding a first-order low pass filter and weights to the exogenous input p, and weights to the uncertainty input and output. The solution algorithm was based directly on those in [238].

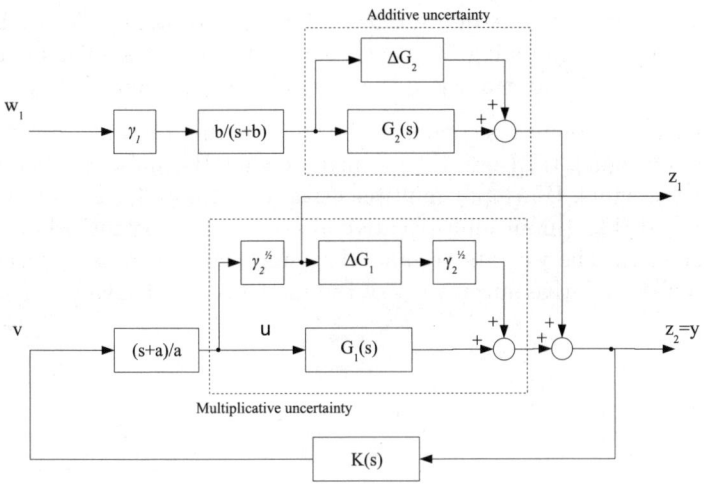

Fig. 4.10. H_∞ example – Adapted from Kienitz et al [242]

- Parker et al [207] used a detailed nonlinear patient model in the development of a continuous-time H_∞ controller, in contrast to Kienitz et al who synthesize the controller from an empirical low-order linear patient model. A non-linear pharmacokinetic/pharmacodynamic model (Sorensen) was first linearised via Jacobian linearisation and then reduced to a third-order linear form by balanced truncation. The reduced linear model was given in state-space form:

$$\dot{x}_{red} = A_{red}x_{red} + B_{red}u + B_m m_d$$
$$y = C_{red}x_{red}$$

and could be split into a process model (G), and a disturbance model (G_m) representing the effects of the meal disturbance (m_d), in similar fashion to those in Kienitz et al (above).

Quoted from [207], "Uncertainty due to differences between an actual patient and the diabetic patient model was thought to be related to variations in model parameters." Sets of three parameters from a possible of eight physiological parameters – five parameters from liver, and three parameters from the peripheral (muscle/fat and tissues) – were chosen for uncertainty characterization, yielding 56 combinations. With each parameter set varied in five permutations (+max, +1/2 max, no change, -1/2 max, -max) about the nominal value, there are a total of 125 variations possible in each three-parameter set (3^5), or 7000 possible perturbed models (56×3^5). The nonlinear model was linearised around each of the 125 parameter variations for each set. The most sensitive parameter set was identified by summing the multiplicative uncertainty

$$U_{rel}(\omega) = \left| \frac{G_{perturbed}(\omega) - G(\omega)}{G(\omega)} \right|$$

in the frequency range of interest $\omega = \begin{bmatrix} 0.002 & 0.2 \end{bmatrix}$ (rad/min) for each perturbation, and then summing over each parameter set. A similar analysis was performed for the uncertainty in the meal disturbance model (G_m).

The controller design was completed by constructing the weighting functions: W_u (input weight), W_m (meal disturbance weight), W_n (noise weight), W_p (performance weight), W_i (input multiplicative uncertainty for perturbed patient model), and W_{im} (input multiplicative uncertainty for perturbed meal disturbance model). The weights W_i and W_{im} were calculated as the least upper bound on the relative uncertainty of the model respectively.

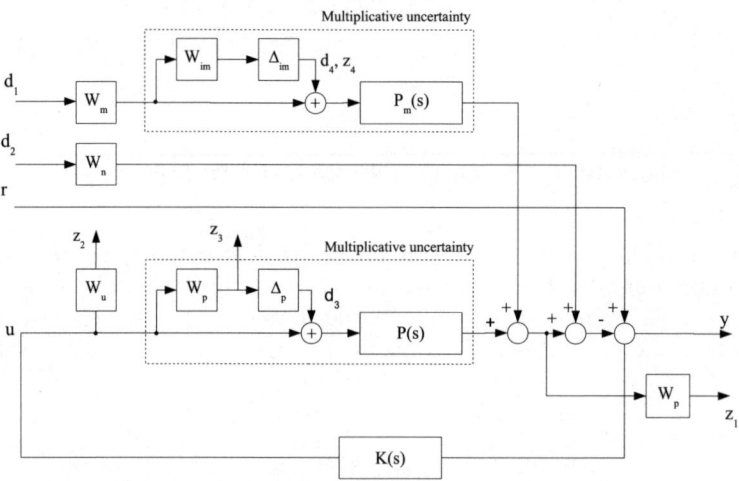

Fig. 4.11. H_∞ example – Adapted from Parker et al [207]

- Ruiz-Velázquez et al [208] had focused on the synthesis of a controller to "compute the intravenous infusion from the subcutaneous measurements of BG concentration such that the glucose level tracks the curve of non-diabetic subjects by accounting". A non-linear pharmacokinetic/ pharmacodynamic compartmental model (Sorensen) was linearised around euglycaemic condition (at 80 mg/dl, with insulin at a constant rate of 22 mU/min). The linearisation of the nonlinear system provided a nominal model around nominal parameters values. The resulting generalized plant $P(s)$ is then obtained by balanced truncation:

$$\begin{bmatrix} z_1 \\ z_2 \\ y \end{bmatrix} = P(s) \begin{bmatrix} d_1 \\ d_2 \\ u \end{bmatrix} = \begin{bmatrix} W_p G_m W_m & 0 & W_p G \\ 0 & 0 & W_u \\ -W_m G_m & -W_n & -G \end{bmatrix} \begin{bmatrix} d_1 \\ d_2 \\ u \end{bmatrix}$$

where G = transfer function of the reduced linear model; G_m= transfer function of the reduced linear model for the meal; W_p = performance weight, chosen such that the frequency content of G_m is captured; W_m = the effect of the meal model; W_n = sensor noise effect; W_u= weight for the control input; d_1 = meal input; and d_2= sensor noise input.

In designing a controller robust for BG regulation, the uncertainty description was computed when parameter variations were incorporated into the nonlinear model. A family of plants was obtained by adding parametric variations into the non-linear model, and then every one of the non-linear models was linearised at the same euglycaemic condition. A multiplicative uncertainty was used:

$$\widehat{G} = [I + W_{ip}\Delta]\,G$$

where $\|\Delta\|_\infty \leq 1$, G is the nominal plant, and W_{ip} is a weighting function. The transfer function W_{ip} was identified via the maximum in the frequency response of the set of multiplicative uncertainty on the nominal plant G:

$$W_{ip}\Delta = \left|\frac{\widehat{G} - G}{G}\right| \qquad \Rightarrow \qquad W_{ip}(j\omega) = \max\left|\frac{\widehat{G}(j\omega) - G(j\omega)}{G(j\omega)}\right|$$

The same procedure was followed to identify a multiplicative uncertainty for the input disturbance (meal), W_{im}. The generalised plant, $P(s)$ became a fifth-order model:

$$\begin{bmatrix} z_1 \\ z_2 \\ z_3 \\ z_4 \\ y \end{bmatrix} = P(s)\begin{bmatrix} d_1 \\ d_2 \\ d_3 \\ d_4 \\ u \end{bmatrix} = \begin{bmatrix} W_m G_m W_p & 0 & W_n G & G_m W_p & W_p G \\ 0 & 0 & 0 & 0 & W_u \\ 0 & 0 & 0 & 0 & W_{ip} \\ W_m W_{im} & 0 & 0 & 0 & 0 \\ -W_m G_m & -W_n & -G & -G_m & -G \end{bmatrix}\begin{bmatrix} d_1 \\ d_2 \\ d_3 \\ d_4 \\ u \end{bmatrix}$$

The synthesis problem is then to find a controller that minimises $z = [z_1, z_2, z_3, z_4, y]^T$ in the H_∞ sense by the classical technique and LMI technique.

- Chee et al [243] designed a "switching controller" based on a non-standard H_∞ control theory [244–246]. The internal model is based on the Minimal model for diabetic state, but linearised at a nominal glucose value. The internal model has the general structure:

$$x(t) = A(t)x(t) + B_1(t)\xi(t) + B_2(t)u(t)$$
$$z_c(t) = K_c(t)x(t), \qquad t \neq iT$$
$$z_d(iT) = K_d(iT)x(iT), \qquad i = 0, 1, 2, \dots$$
$$y(iT) = C_i x(iT) + v_i, \qquad i = 0, 1, 2, \dots$$

where $x(t)$ = state vector; $\xi(t)$ = external glucose disturbance input; $u(t) = \{0, U_0, U_1, \dots\}$ = a sequence of (pre-defined) insulin infusion rate; $z_c(t)$ = continuous controlled output; $z_d(iT)$ = discrete controlled output; $y(iT)$ = discrete BG measurement; v_i = noise. The control strategy is a rule for switching

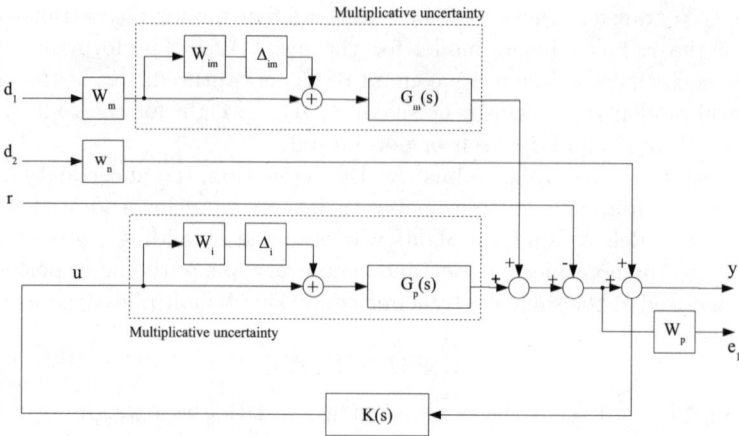

Fig. 4.12. H_∞ example – Adapted from Ruiz-Velázquez et al [208]

from one sequences of insulin rates to another. A performance index similar to that of equation 4.12 is used, and the related Riccati equation is derived using the standard H_∞ method initially. Bertsekas et al's method [247] is then used to transformed the performance index into a form that does not depend on $\xi(t)$. To find the optimum insulin infusion sequence, dynamic programming is used to calculate the performance index associated with each possibilities of $u(t)$ sequence with arbitrary $y(iT)$.

4.5.6 Note

It should be noted that most control algorithms described above use BG level measured at i.v. catheter or subcutaneous regions. Insulin are delivered either intravenously or subcutaneously, and not directly at the portal circulation. This is partly because it is safer to access blood vessels at the peripheral and subcutaneous site, than at the portal site, despite the fact that insulin delivered at peripheral site usually causes hyperinsulinemia, and adverse effects have been reported (e.g. an increased risk of developing Alzheimer disease [154, 155]).

4.6 Conclusion

The design of effective control algorithms require a knowledge of the characteristics of the individual components that constitute the loop (i.e. the sensor, insulin dynamics etc). Furthermore, practical implementation also needs to address the issue of noise in measurement and uncertainties in the model. The use of full and comprehensive glucose-insulin description requires a knowledge of all the interdependent physiological processes involved, and also on the accessibility of the parameters involved. Some parameters are not easily obtainable with simple tests, and may not be measurable due to technical barriers, ethical issues, or

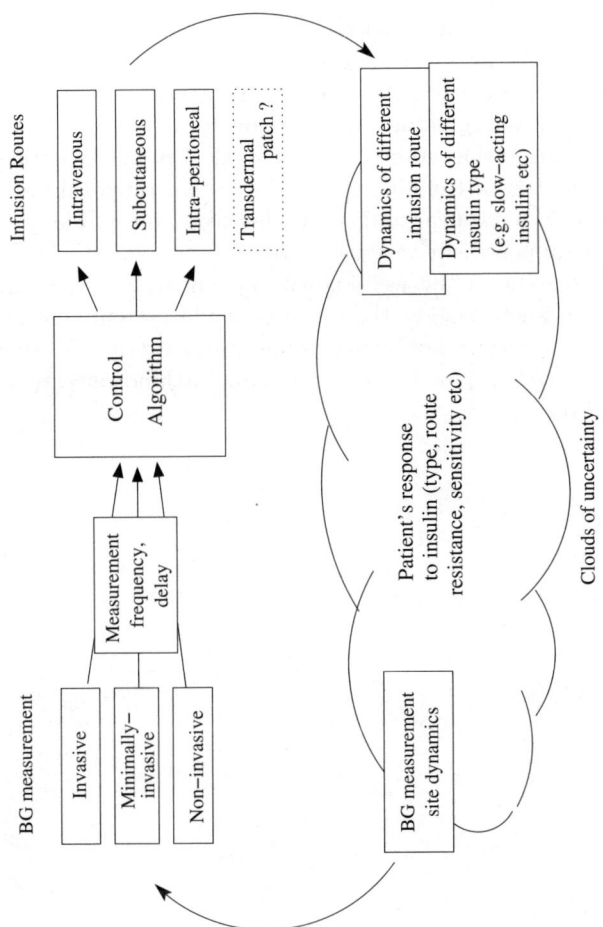

Fig. 4.13. Map of inter-relationship in a closed-loop insulin delivery system for BG control

cost-benefit issues. Having said, there has been effort in developing algorithms that uses "average" model parameter values as a start, and then "personalized" as data are gathered (e.g. Hovorka et al).

The application of mathematical modelling and control theory certainly helped in the understanding of the glucose-insulin kinetic and the subsequent control problem, bringing us one-step closer to achieving better glucose regulation. What remains missing is a reliable, accurate and minimally/non-invasive glucose sensor.

4.7 Summary

This chapter examines the two approaches to glucose controller design. Model-less approach relies heavily on rules or empirical method to devise the control algorithm. In contrast, model-based approach attempts to understand, and mathematically "capture" the underlying system, before devising the control strategy. When the underlying system is sufficiently captured in the form of mathematical equations, glucose-control problem can be viewed as mathematical problem, and solved using standard/state-of-the-art mathematical techniques. Mathematical models of the glucose-insulin system ranges from simpler linear models, through to non-linear and comprehensive models. Many control strategies have been proposed, in various degrees of complexity. This chapter then briefly described some of the well-known implementations as found in the literature. To-date, knowledge of the glucose-insulin system continues to evolve, and new control strategies are being proposed constantly. The unavailability of a reliable and accurate glucose sensor remains a hindrance to the solving of glucose-control problem.

Closed-Loop Control Apparatus Example

MiniMed® CGMS is, perhaps, the first commercially-available subcutaneous glucose sensor for use by the masses. CGMS consists of a disposable subcutaneous glucose sensor assembly connected by a cable to a pager-sized glucose monitor. CGMS takes a glucose measurement every 10 seconds and stores a smoothed and filtered average of these values every 5 minutes in the memory bank on-board.

To view the BG level readings, the sensor signal collected by CGMS must first be downloaded to a PC using MiniMed® Com-Station via RS232-C serial communication (Fig 5.1).

In this chapter, we show how to:

- Construct a bedside closed-loop glucose control apparatus (with a basic control algorithm) using CGMS and other off-the-shelf components;
- Create a portable closed-loop system by pairing the subcutaneous glucose sensor assembly with a syringe pump and microprocessor.

5.1 CGMS Integration

With PC interface included in CGMS, it is possible to connect CGMS to a personal computer (PC), and adding a bedside infusion pump to effect a closed-loop glucose control apparatus. In this section, a "basic" bedside closed-loop insulin infusion system is described. Although the system uses intravenous insulin infusion, it is not difficult to expand the work to use subcutaneous infusion pump.

5.1.1 Hardware Integration

The three components of the closed-loop insulin delivery system are:

- personal computer, running an operating system of your choice;
- MiniMed-Medtronic CGMS connected to the PC through RS232-C serial cable;
- Computer-controllable medical infusion pump.

F. Chee & T. Fernando: Closed-Loop Control of Blood Glucose, LNCIS 368, pp. 109–126, 2007.
springerlink.com   © Springer-Verlag Berlin Heidelberg 2007

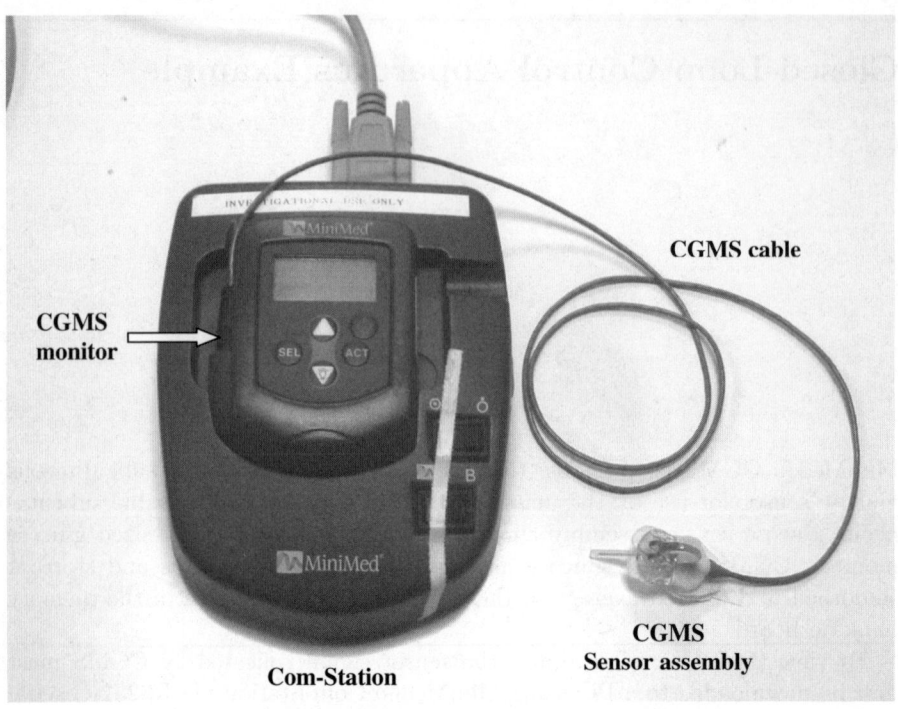

Fig. 5.1. CGMS hardware

As a real-life example, we used a PC with Intel® Pentium III 800MHz running Microsoft® Windows® 98SE (which was popular in year 2000), and an IMED PC-1 peristaltic pump (Alaris Medical System, San Diego CA) that can be controlled from a PC through RS232-C serial null-modem cable (Fig 5.2).

5.1.2 Software Integration

Once the hardware is in place, the major functions of the closed-loop system are managed by software. The software performs four logical functions:

- Sensor data collection
- Pump driving
- Control algorithm computing
- Alert reporting (alarming)

Sensor Data Collection

CGMS is approved for investigational use by the FDA (Food and Drug Administration, USA), but is not designed to give real-time glucose information (at the time of writing). Each sensor is meant to be worn for up to three days, with the patient performing SMBG (i.e. Self-Monitoring of Blood Glucose) at least four times a day and entering the SMBG result into the monitor. When initiated, MiniMed® Com-Station Software V1.6 downloads the "raw data" (i.e. current, voltage from the sensor assembly) that are stored onboard CGMS, translating it and then storing it as plain text data file on the PC. MiniMed's Solution Software V1.1A then formats this text data file, and applies Regression Calibration to estimate BG readings obtained during the three days retrospectively.

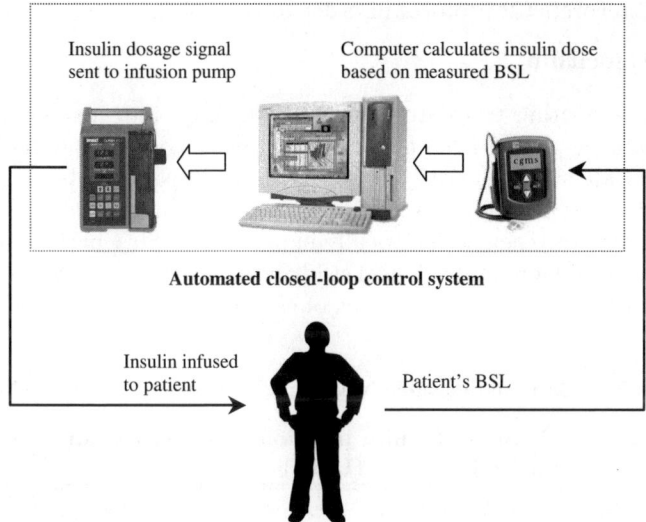

Fig. 5.2. Closed-loop insulin delivery system components

To use CGMS in real-time for closed-loop control purposes would required:

- sensor "raw data" readings to be (automatically) downloaded every 5 min, and processed without using MiniMed's Solution Software V1.1A;

  This can be achieved by simulating the mouse-clicks and keyboard key presses using software.

- a modification to the regression calibration method to allow real-time estimation of BG.

  Regression calibration was implemented in MiniMed's Solution Software V1.1A as an off-line method, whereby BG readings were analysed in a retrospective manner (see Section 2.5.2). To obtain BG readings on-line, a quick way would be to use multi-point in-vivo calibration, but with the calibration conducted in real-time.

Regression calibration was applied whenever the sensor readings deviated from the reference glucometer BG reading by a pre-determined amount (e.g. 3 mmol/l). The glucometer value was taken as the calibration BG entry.

Pump Driving

The IMED Gemini PC-1 peristaltic pump uses a C2 Communication Protocol (Alaris Medical System, San Diego CA). C2 Protocol defines how communications can be achieved between a computer and a pump through a RS232-C serial interface by means of pre-defined command set. The software needs to generates/interprets C2 protocol in order to control the pump.

Control Algorithm

The control algorithm translates BG levels readings into matching insulin delivery rates to be given to the patients. As a basic control apparatus, a sliding table control algorithm is used to serve as an example. The basic sliding table has the form shown in Table 5.1, and features a commonly used starting scale for treating hyperglycaemia in critically ill patients at Sir Charles Gairdner Hospital, Perth (where a real-life clinical study was performed using the apparatus). Insulin delivery is adjusted at the turn of each hour, and maintained at the new

Table 5.1. Basic sliding table, with a region assigned to each BG ranges

BG ranges (mmol/l)	Insulin infusion rate (U/hr)	BG Region
>20.0	4	4
15.1 – 20.0	3	3
10.1 – 15.0	2	2
6.1 – 10.0	1	1
0 – 6.0	0	0

delivery rate during the hour. In conjunction with the basic sliding table control algorithm, the following extra steps were included:

1. Ability to start at an insulin rate higher or lower than those set-out in Table 5.1, by adding a constant U_{offset} that adds to the basic set of infusion rates.

   This is to address the issues that not all patients have the same insulin requirement as exemplified in Table 5.1. With any selected "starting" infusion rate, the initial U_{offset} would be automatically calculated by

$$U_{\text{offset}} = (\text{first prescribed rate} - \text{BG region})^+$$

   where the operation $(\cdot)^+$ would return a "-1" if the enclosed term is negative-valued. Any subsequent insulin will be given according to

$$U_{\text{given}} = \text{BG region} + U_{\text{offset}}$$

2. Automatic scale-shift every 6 hours.

   This is to address the issues that any static sliding table may not be appropriate if a patient's medical condition changed during the course of treatment. The scale-shift procedure alters the value of U_{offset} in the following manner:

   - When there was an upward region transition:
     $$U_{\text{offset}}(n) = \begin{cases} U_{\text{offset}}(n-1) + 1, & \text{All other upward transition} \\ U_{\text{offset}}(n-1), & \text{From Region 0 to Region 1} \end{cases}$$

   - When there was no change in BG region:
     $$U_{\text{offset}}(n) = \begin{cases} U_{\text{offset}}(n-1) + 1, & \text{Region 2 and above} \\ U_{\text{offset}}(n-1), & \text{Region 1} \\ U_{\text{offset}}(n-1) - 1, & \text{Region 0} \end{cases}$$

   - When there was a downward region transition:
     $$U_{\text{offset}}(n) = \begin{cases} U_{\text{offset}}(n-1), & \text{All other downward transition} \\ U_{\text{offset}}(n-1) - 1, & \text{From Region 1 to Region 0} \end{cases}$$

3. Rules to immediately reduce, or turn-off insulin delivery when BG fell below a pre-defined threshold (rather than waiting until the start of the next hour).

   This is to safeguard against unwanted hypoglycaemic episodes and protect the patients. The rules are:

   - On detection of BG rise or fall into 6–7 mmol/l range, adjust insulin rate as per U_{given} immediately;
   - On detection of BG falling below 5.5 mmol/l, set $U_{\text{given}} = 0$ and $U_{\text{offset}} = 0$ immediately.

Alert Reporting

The software also needs to report any errors or alarms to the user. This includes any errors in sensor reading, pump driving or other conditions.

5.1.3 Clinical Trial Example

To show that such a control apparatus actually works, we included a few clinical results obtained during a clinical study on critically-ill patients (see [167]). Note that the study was approved by the Institutional Ethics Committee of Sir Charles Gairdner Hospital, Perth, Western Australia, and the Therapeutics Goods Administration (TGA), Canberra, Australia. Informed consent were obtained from the patients or their legal surrogates prior to conducting the study.

Fig 5.3 shows the best performing glucose sensor from the clinical study. The sensor BG reading tracked the MediSense® 2 glucometer reading with 2.9±9.6% error (mean±2SD). With such excellent BG readings, the action of the sliding table control algorithm can be examined. One can see the sliding table slowly increased the insulin rate to bring BG to near 10 mmol/l at the end of the 24-hour period. This result showed the classical control performed by sliding table

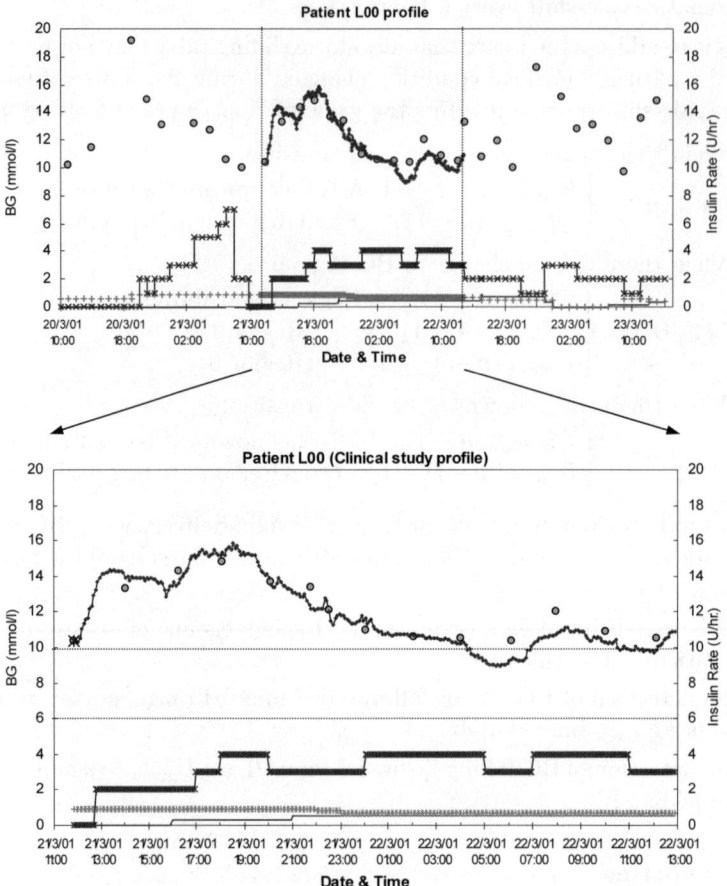

Fig. 5.3. Profile for Patient L00. Shown are the glucometer reading (○), sensor reading (●), insulin infusion rate (×), Total Parenteral Nutrition (+, in ml/hr), naso-gastric feed ("−", Jevity® in ml/hr), and sensor calibration (✳). Jevity® is a nasogastric feed from Abbott Laboratories Inc, USA, and contains 36.5 g of carbohydrate per 237 ml (8FL Oz).

method. The average BG attained during the trial period was 12.0±3.2 mmol/l (mean±2SD). The control algorithm was configured to bring BG to the range 6–10 mmol/l, taking a conservative control approach. Fig 5.4 shows the complete closed-loop infusion system used in the clinical study.

It was observed that the glucose sensor did not always perform as expected. Some observed sensor performances that are undesirable for use in closed-loop control included below:

1. Sensor signals (Isig in nA) decreases with time (Fig 5.5):
2. Sensor signals (Isig in nA) delayed by 1 hour with respect to the glucometer readings (Fig 5.6):

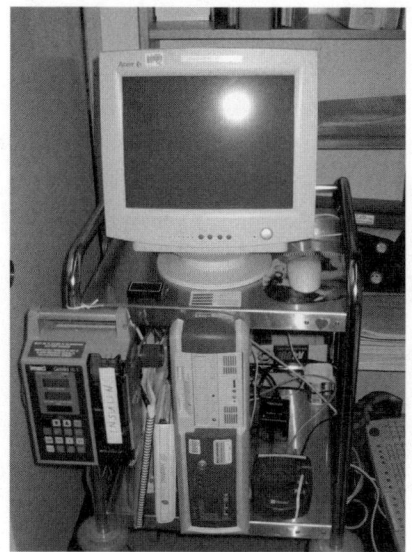
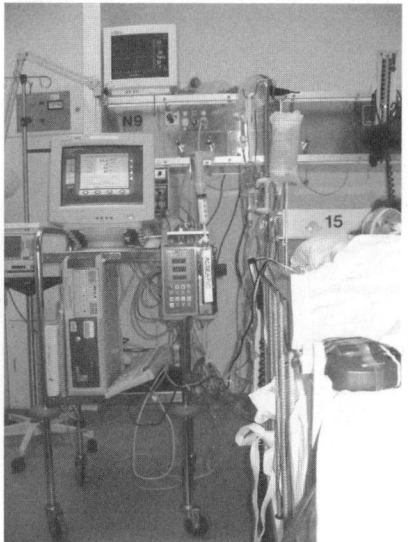

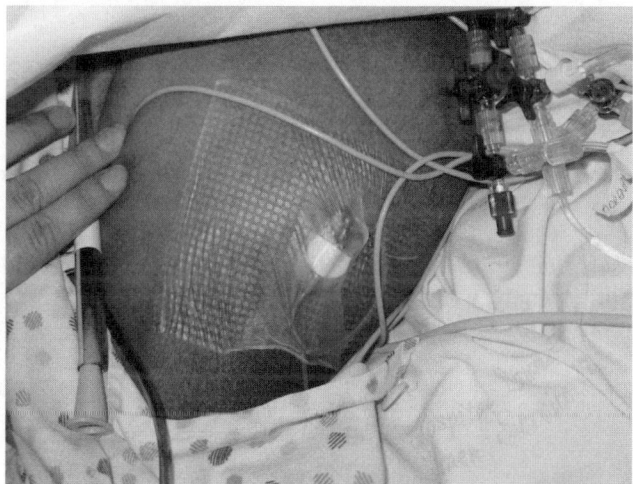

Fig. 5.4. Photographs of the complete system used during clinical studies (top). CGMS glucose sensor on patient's upperarm area (bottom).

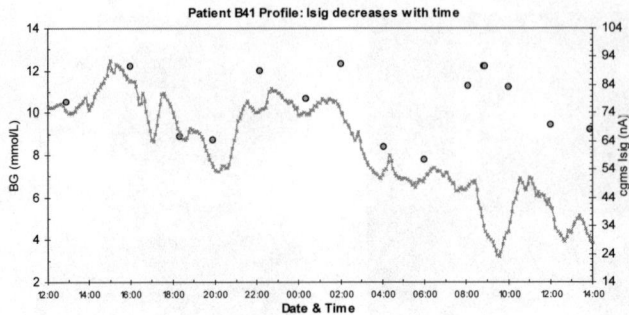

Fig. 5.5. Sensor Isig decreases with time, when compared to the glucometer readings (∘)

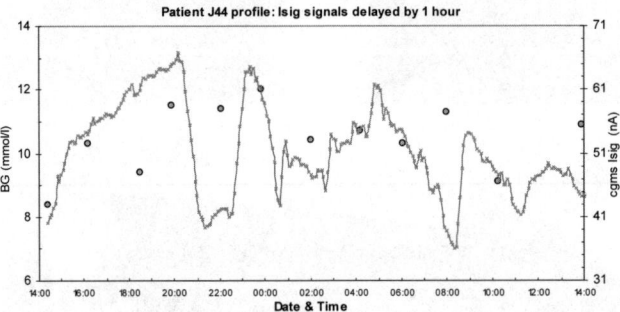

Fig. 5.6. Sensor Isig delayed by 1 hour, with respect to the glucometer readings (∘)

3. Sensor signals (Isig in nA) fluctuating around the glucometer readings, while decreasing in strength with time (Fig 5.7):

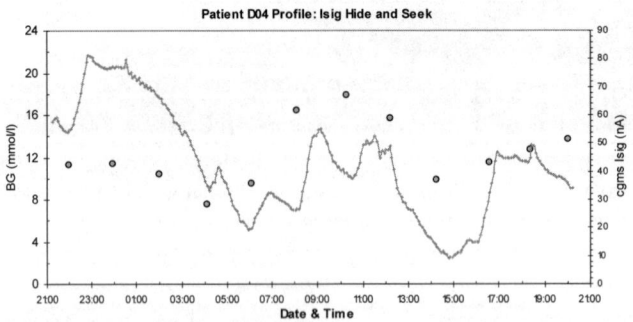

Fig. 5.7. Undulating sensor Isig when compared to the glucometer readings (∘), while decreasing in strength with time

Considering the many factors that can influence the sensor-glucose dynamics in the subcutaneous space, it may not be surprising that the sensor signal fluctuated in all the results obtained. In addition, the clinical study was conducted on the critically-ill patient population, which has a different physiological condition compared to healthy individuals. The estimation of BG from the sensor current signal is dependent on three factors:

1. the linearity between sensor signal and ISF glucose concentration (interaction factor).
2. the linearity between the ISF glucose concentration and blood glucose concentration.
3. the calibration factors that translate the current signals into BG estimation.

Among the five clinical studies, Patient L00 (Fig 5.3) exhibited the least changes in sensor linearity, with the highest change at 0.17 mg/dl per nA, which amounted to 1.7–17 mg/dl (or 0.09–0.9 mmol/l) for a current range of 10–100 nA.

The example of the bedside system shown in this section showed that a bedside closed-loop system can be constructed fairly easily, due to the commercial-availability of the components – especially the glucose sensor.

5.2 Miniaturisation – Glucose Sensing Circuit

In this section, we show how to integrate the disposable subcutaneous glucose sensor assembly with a "glucose sensing circuit". The glucose sensing circuit interfaces physically with the CGMS glucose sensor assembly (Fig 5.1) through the CGMS cable, and provides all the necessary electrical environments required for glucose sensing. To create a portable closed-loop system, the glucose-sensing circuit is further paired up with a syringe pump and a microprocessor.

This integration is done purely to illustrate how a commercial glucose sensor can be used for research and educational purposes. We (as well as the FDA (Food and Drug Administration, USA)) are, in no way, approving this modified arrangement for use in human subjects.

5.2.1 Background

CGMS uses amperometry as its main measuring technique (see Section 2.1). The commonly used sensor configurations are:

1. two-electrode system — a platinum (Pt) working-indicating electrode with a Ag/AgCl reference-counter electrode, such as those described in [16,248,249].
2. three-electrode system — a platinum (Pt) working electrode, an Ag/AgCl reference electrode, and a platinum (Pt) counter electrode such as those described in [33,248,250,251].

CGMS uses a three-electrode system. A circuit that can be used with CGMS is presented here, and it consists of three components:

1. a constant potential application circuit.
2. a nano-ampere measuring circuit.
3. a supply splitting circuit.

5.2.2 Constant Potential Application Circuit

The constant potential application circuit (formed by op-amps 2 and 3 in Fig 5.8) functions to impose a constant potential between two electrodes, for the hydrolysis of hydrogen peroxide (see Table 2.1 in Chapter 2).

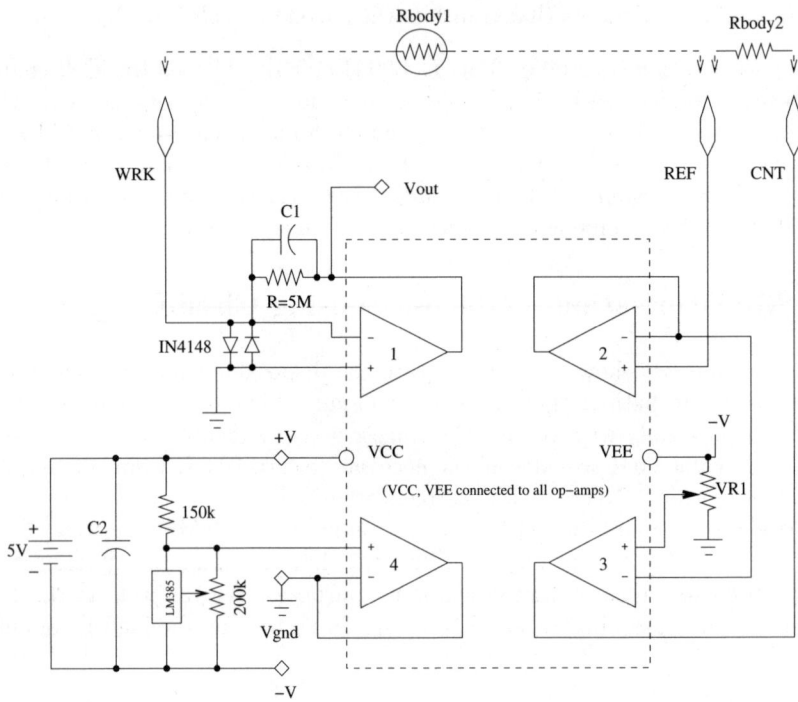

Fig. 5.8. Glucose sensing circuit for integration with MiniMed® CGMS glucose sensor assembly. Figure shows a constant potential application circuit (right), a nano-ampere measuring circuit (upper left) and a supply potential splitting circuit (bottom left). The resistances, R_{body1} and R_{body2} represent the flow of the current generated from the hydrolysis of H_2O_2, across the constant potential applied to the electrodes.

In this design, a negative potential is applied at the reference electrode, *REF*, with respect to the working electrode, *WRK*. The design used two op-amps (i.e. op-amps 2 and 3) arranged in voltage-follower configuration, and the variable resistor VR1 tuned to provide the negative potential. Since there should be no currents flowing into the input terminals of op-amps 2 or 3 (i.e. due to the high resistance of the terminals), any current generated from the hydrolysis of H_2O_2

would flow to the counter electrode, *CNT*, and then sink to VEE via the output terminal of op-amp 3. This configuration allows the voltage at REF to remain constant.

Fig. 5.9. Connection to the CGMS sensor assembly (Source: MiniMed's patent WO00/74753A1, US6,360,888B1)

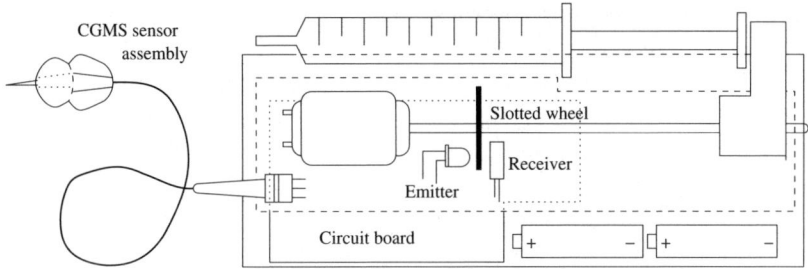

Fig. 5.10. Illustrative diagram of pump assembly

5.2.3 Nano-ampere Measuring Circuit

The generated current, due to its small magnitude (in nano-ampere range), is measured by a nano-ampere measuring circuit, which is formed around op-amp 1. The circuit employs a *transimpedance* topology, and converts the out-flowing current (with respect to the op-amp's "-" input) into a positive output voltage. The output voltage is proportional to the current, and is thus proportional to the amount of glucose present at the sensor insertion site. The two diodes that are connected to the op-amps inputs function to limit excessive current from damaging the op-amp. The capacitor in series with the $5M\Omega$ resistor forms a low-pass filter with a cut-off frequency of $1/\sqrt{2\pi \cdot (5M\Omega \times C1)}$ to eliminate noise in the measurement.

5.2.4 Supply Potential Splitting Circuit

To permit the use of a single 5V battery, and yet simultaneously allowing the existence of a negative potential (for the reference electrode), and a positive voltage output (at the nano-ampere measuring circuit), a supply potential splitting circuit is used. Formed by op-amp 4, the supply splitting circuit selects a potential within the source potential range (by means of a LM385 adjustable micropower voltage reference diode feeding into op-amp 4), and maps it for use as "ground" reference by the other circuits. This effectively splits the source potential into two separate potential sources, giving a positive and negative voltage reference. This eliminated the need for a separate generation of a negative power supply.

The use of op-amps with very low power consumption minimised the effect of limited current sourcing/sinking ability of the op-amps configuration employed in the supply splitting circuit.

Used in conjunction with the CGMS sensor assembly and cable, this glucose sensing circuit allows minute-by-minute measurement of subcutaneous glucose concentration, to be used by a control algorithm in the insulin delivery device.

5.3 Miniaturisation – Syringe Pump and Microprocessor

The glucose-sensing circuit can be interfaced with a microprocessor and syringe pump to effect a closed-loop control system.

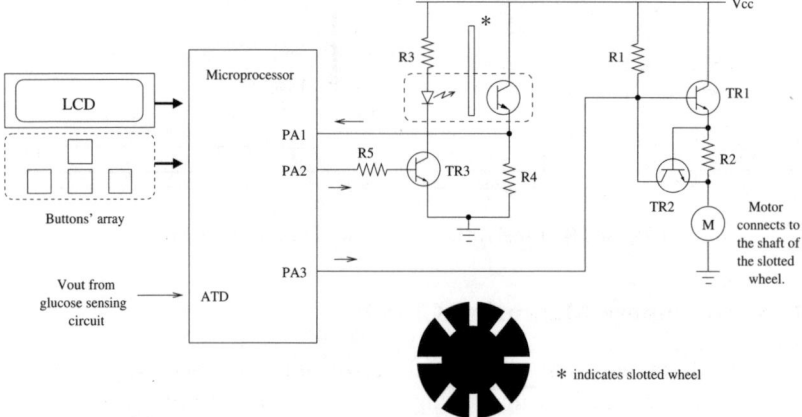

Fig. 5.11. Syringe pump circuit schematic, showing the interface between the motor driving circuit, the motor rotation detector circuit, with the microprocessor

The use of a microprocessor allowed flexibility in changing functionality (in the control algorithm). Fig 5.11 shows the circuit design of the microprocessor-controlled syringe pump, which consists of

1. a motor driving circuit
2. a motor rotation detector circuit
3. a liquid crystal display (LCD) and hardware buttons
4. a microprocessor

5.3.1 Motor Driving Circuit

The motor driving circuit consists of a voltage source with current limiter. The motor is turned on by a logic high placed on the I/O pins of the microprocessor that connects to transistor TR1 (i.e. PA3 in Fig 5.11). When the motor drew more than a pre-defined current determined by the value of resistor R2 (e.g. when there is a flow resistance in the delivery catheter), TR2 would turn on, limiting the current to the motor, and protecting the motor windings from overcurrent.

5.3.2 Motor Rotation Detector Circuit

The motor rotation detector circuit consists of a light emitting diode, and a phototransistor, which can also be replaced by a slotted optical switch assembly. A slotted wheel, which is mounted on the motor shaft, is located between the slot of the optical switch assembly. Feedback on the degree of motor rotation was provided by the motion of the wheel interrupting the light beam between the slot (see Fig 5.10 and Fig 5.11).

As the optical switch is mainly used to detect the degree of rotation made by the motor, the microprocessor can be programmed to save power by switching on the optical switch only when the motor is turned on. The motor and the optical switch can then be switched off when the motor has rotated a pre-selected amount, or when a time-out has occurred. The values of resistor R1 to R4 vary with different types of motor and optical switch desired.

5.3.3 LCD and Hardware Buttons

The LCD and hardware button are provided for sensor value readout and infusion rate programming or viewing. The designer can also choose to provide dials in place of the LCD and hardware buttons.

5.3.4 Microprocessor

A microprocessor is used to coordinate the on-off sequence of the syringe pump motor mechanism, to digitise and perform signal processing on the sensor readings, and to allow the user interaction with the system (e.g. checking the sensor readings, or changing the infusion rates via the LCD and hardware buttons). Sensor signals are digitized using the on-board Analog-To-Digital (ATD) channel. As described, the bulk functionality of the system are software-oriented.

5.4 Signal Processing and Control Algorithm Programming

Noise will be inevitable when sensor signals are in nano-current magnitude, and hence filtering of noise was required. Since the sensor signal is fed into the microprocessor, most digital filters could be implemented.

In this design, a simple moving average filtering was employed to obtain a relatively stable current readout on the microprocessor. A fast filtering mechanism was required to permit real-time reporting of sensor readings. Table 5.2 describes a fast moving average algorithm employed in the system. The algorithm used a first-in first-out buffer, with 1000 taps set aside for the averaging process.

Table 5.2. Fast moving average filter

The moving average filter has the following transfer function:

$$G(z^{-1}) = \frac{Y(z^{-1})}{X(z^{-1})}$$

$$= \frac{1 + z^{-1} + \cdots + z^{-M+1}}{M}$$

where $X(.)$ and $Y(.)$ represent the input and output of the filter respectively, and M is the length of the averaging window.

The above filter can be implemented in software as follows:

$$y[n] = \frac{x[n] + \cdots + x[n - M + 1]}{M} \tag{5.1}$$

or

$$y[n] = y[n - 1] + \frac{x[n]}{M} - \frac{x[n - M]}{M} \tag{5.2}$$

The number of operations required to compute $y[n]$ in equation (5.2) is less than the operations required in equation (5.1).

5.5 Circuit Performance

To examine the performance of the nano-ampere measuring circuit, used in conjunction with the fast moving averaging algorithm, the linearity of the measuring circuit was tested by injecting test input currents into the circuit and measuring the output of the circuit. Test currents in the nano-ampere (nA) range were generated, by means of using different resistor values for R_{body1} and R_{body2} to effect different current values (see Fig 5.8). Since R_{body1} and R_{body2} reflect the generation of current from hydrogen peroxide and its flow across a constant potential, this current generation can be simulated by using different resistors.

For the purpose of the test, R_{body2} was set equal to R_{body1}, and the glucose sensing circuit was constructed using the National Semiconductors LMC6464 op-amp. This micro-power quad CMOS operational amplifier features a 20μA supply current per amplifier, >10TΩ input resistance, 150fA input current, and +5V DC single supply operation.

Fig 5.12 shows a comparison of the output measured by the nano-ampere measuring circuit (i.e. V_{out}/R in Fig 5.8) with the test currents applied at the input of the circuit (i.e. at *WRK* electrode). Both the input current and V_{out} were measured using MetraHit 30M (Gossen-Metawatt GmbH, Germany) digital multimeter which has internal noise-filtering circuit for stable readings. The result showed that the output measurement is linear with respect to the input current.

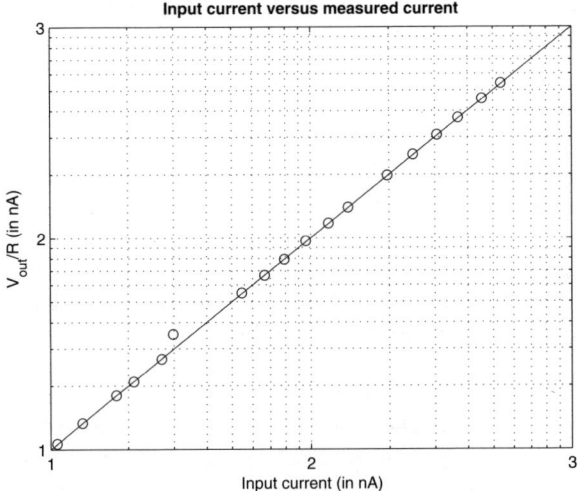

Fig. 5.12. The output measured by the nano-ampere measuring circuit is linear with respect to the applied test current, over the tested range of 10 nA to 400 nA

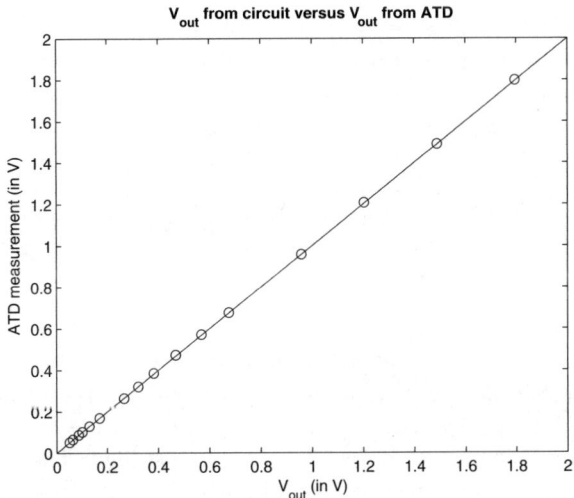

Fig. 5.13. V_{out} is linear with respect to the ATD's measured output (10-bit ATD)

To examine the performance of the moving averaging filter in combination with the ATD, the output of the nano-ampere measuring circuit was compared to the those sampled by a 10-bit ATD channel and subsequently filtered by the moving average algorithm. Fig 5.13 showed that V_{out} is linear with respect to the ATD's output measurement.

5.6 Basic Safety Issues

The microprocessor-controlled closed-loop system discussed in this chapter is in its basic functional form and is not suitable for use in any testing involving live human subjects. The design here provides a ready reference for the construction of a compact and portable closed-loop system. The system hardware design would still need to undergo consideration of the issue of safety. Some possible safety issues (and guidelines to overcome them) are:

- Occlusion – obstruction in the tubings preventing the flow of infusate.

  Occlusion can happen at the "patient" side (i.e. connection between the tip of the tubing and the patient's vessels or skin) or at the "pump" side (i.e. connection between the syringe and the tubing). Occlusion can be sensed from the lack of rotation of the pump motor, and the software can be made to detect this event. However, for added safety, a dedicated pressure sensor should be attached to the tubings to increase the detection sensitivity.

- Circuit failure, e.g. motor rotation detector circuit failure.

  The software should be written such that any failure in the motor rotation detector or motor driving circuit would result in an immediate shut-down of the infusion, and an alert to be sent to the user.

- Low battery alarm.

  The fitting of a low battery alarm is useful. This can either be implemented in software (for certain types of microprocessors) or by installing an external battery monitoring circuit.

- Air-in-line detection.

  In the case of intravenous delivery, the presence of air bubbles in the tubing would be undesirable (and sometimes disastrous). Care should be taken to emptied the bubbles from the tubing before connecting to the patient. Alternatively, an "air-in-line" detector could be installed. The microprocessor-based syringe pump designed in this did not include an "air-in-line" detector, giving other researchers the flexibility to chose their detection method (e.g. ultrasonic detector).

Fig 5.14 shows a working example of the system built around an existing commercially-available syringe pump (SIMS Graseby® MS-26 Syringe Driver, Graseby Medical Ltd, Watford, Hertfordshire, UK), as an alternative to a custom-made pump.

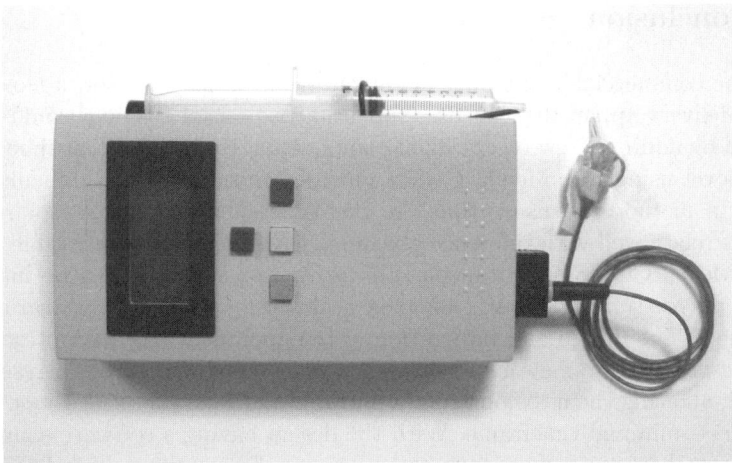

Fig. 5.14. Working example of a portable closed-loop insulin delivery apparatus

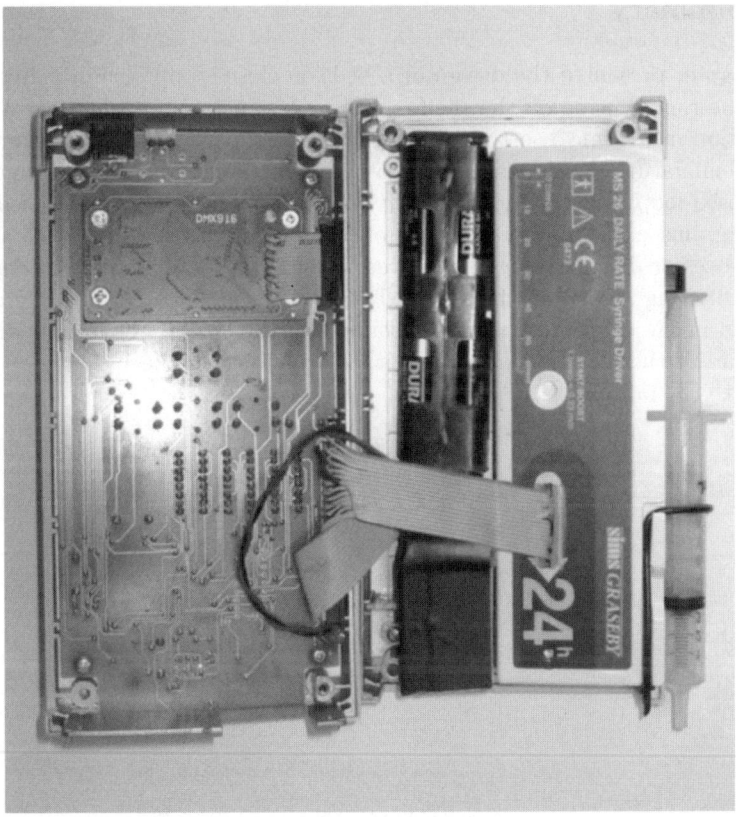

Fig. 5.15. The internal of the example portable closed-loop insulin delivery apparatus

5.7 Conclusion

With the commercial-availability of subcutaneous glucose sensor, a closed-loop insulin delivery apparatus that uses subcutaneous sensor readings could be constructed by adding a few off-the-shelf components. However, a real-time control of BG level using MiniMed® CGMS glucose sensor would be difficult unless there is a method to ensure that the real-time subcutaneous sensor readings could correlate well with reference glucometer readings in a predictable way.

MiniMed® CGMS subcutaneous glucose sensor assembly can be interfaced with a glucose-sensing circuit, allowing a closed-loop control apparatus to be pocketable. The design and integration of the apparatus has been presented as a basic "working" framework to provide a ready reference to other research in the field, allowing them to construct a portable and wearable closed-loop control system in a minimal timeframe. With the design having a software component, other researchers can conveniently tailor aspects of the system according to their needs.

5.8 Summary

This chapter presented the design of a bedside closed-loop insulin delivery system, constructed using off-the-shelf components. Such an apparatus was shown to function in a clinical settings under constant supervision. As a natural step towards miniaturisation, this chapter detailed the design of a portable and wearable closed-loop delivery system that uses CGMS as its glucose sensor. Circuit diagrams explaining the design of the glucose sensor interface, and the microprocessor-driven syringe pump were presented. The chapter then described a noise filtering algorithm that worked well with the glucose sensing circuit introduced in the chapter. A working example of the design was presented in the later part of the chapter, together with discussions on the basic safety issues related to the pump.

6

Conclusions

Researches have found that a well-controlled BG greatly reduces the mortality and morbidity associated with diabetes. A closed-loop control system could be used to achieve a tight BG control. It can be seen (in Chapter 4) that designing a closed-loop system for BG control is not an easy task. The design of the closed-loop system needs to consider the characteristics of each components in the system, the design of the control algorithm, and how the combination (of the sensor, infuser and algorithm) can achieve the desired outcome. The design of the control algorithm depends on the selection of the BG measurement method and insulin infusion routes. The selection of the BG measurement method and insulin infusion route depends on the available technology and issue of safety associated with the selected method/routes.

It is not possible to describe all the mathematical models and control algorithms reported in the literature in one book. This book gives the frequently cited algorithms & models as examples. The readers are encouraged to read other books in parallel, to get a different perspective.

In this book, our aims were:

- to show that blood glucose control is important not only in ambulatory persons, but also in critically-ill patients;
- to introduce the reader to the basics of closed-loop blood glucose control, which included the functional description of the gluco-regulatory system, and the various components making up the complete system;
- to show, as an example, how a commercially available glucose sensor can be used to build a bedside closed-loop system, and how can it be incorporated to form a pocketable microprocessor-controlled closed-loop system;
- to show that glucose control is a vast and complicated field, simply because our body's glucose-regulatory system is complex and inter-related with various parts of the body function.

This book does not discuss biotechnology, gene therapy, stem cells or pancreas/beta-cell transplantation, which also played a part in helping to cure diabetes.

F. Chee & T. Fernando: Closed-Loop Control of Blood Glucose, LNCIS 368, pp. 127–128, 2007.
springerlink.com   © Springer-Verlag Berlin Heidelberg 2007

The ultimate goal in closed-loop control of BG is to have a system that regulates BG using the non-invasive glucose sensing and insulin delivery, and is also capable of being "set and forget". This goal may not be too far away, as progress is constantly being made both in the field of non-invasive glucose sensing and insulin delivery. Perhaps, one day, the current landscape may dramatically change, as new understanding of the glucose-regulatory system emerge, bringing new insight into the way blood glucose will be controlled.

A

Mathematical Derivation

A.1 Solving Recursive Least Square Estimation

Recall in Chapter 4 Section 4.5.3 that the optimal filter coefficients can be found by setting the derivative of $\xi(n)$ with respect to $w_m^*(n)$ to zero to obtain:

$$\frac{\partial \xi(n)}{\partial w^*} = \sum_{k=0}^{n} \lambda^{n-k} e(n,k) \frac{\partial e^*(n,k)}{\partial \mathbf{w}^*}$$

$$\Rightarrow \quad 0 = \sum_{k=0}^{n} \lambda^{n-k} \left(d(k) - \mathbf{w}^T(n)\mathbf{x}(k) \right) \frac{\partial \left(d(k) - \{\mathbf{w}^*\}^T \mathbf{x}^* \right)}{\partial \mathbf{w}^*}$$

$$0 = \sum_{k=0}^{n} \lambda^{n-k} \left(d(k) - \mathbf{w}^T(n)\mathbf{x}(k) \right) \{\mathbf{x}^*\}^T$$

$$\Rightarrow \quad \left(\sum_{k=0}^{n} \lambda^{n-k} \mathbf{x}^*(k)\mathbf{x}^T(k) \right) \mathbf{w}(n) = \sum_{k=0}^{n} \lambda^{n-k} d(k)\mathbf{x}^*$$

i.e.

$$\Phi(n)\mathbf{w}(n) = \theta(n) \tag{A.1}$$

where

$$\Phi(n) = \sum_{k=0}^{n} \lambda^{n-k} \mathbf{x}^*(k)\mathbf{x}^T(k)$$

$$= \left[\sum_{k=0}^{n-1} \lambda^{n-k} \mathbf{x}^*(k)\mathbf{x}^T(k) \right] + \lambda^{n-n} \mathbf{x}^*(n)\mathbf{x}^T(n)$$

$$= \left[\sum_{k=0}^{n-1} \lambda^{(n-1)-k} \mathbf{x}^*(k)\mathbf{x}^T(k) \right] + \mathbf{x}^*(n)\mathbf{x}^T(n)$$

$$= \lambda\Phi(n-1) + \mathbf{x}^*(n)\mathbf{x}^T(n)$$

is of recursive relationship, and

F. Chee & T. Fernando: Closed-Loop Control of Blood Glucose, LNCIS 368, pp. 129–131, 2007.
springerlink.com   © Springer-Verlag Berlin Heidelberg 2007

$$\theta(n) = \sum_{k=0}^{n} \lambda^{n-k} d(k) \mathbf{x}^*$$
$$= \lambda \theta(n-1) + d(n) \mathbf{x}^*(n)$$

where $\mathbf{x}^*$ denotes a complex conjugate.

To solve equation A.1, it is necessary to find the inverse of $\Phi(n)$. This can be achieved by using Matrix Inversion Lemma:

If A and B are $M \times M$ positive definite matrices, D is a $N \times N$ matrix, C is a $M \times N$ matrix which are related by

$$A = B^{-1} + CD^{-1}C^T$$

then

$$A^{-1} = B + BC(D + C^T BC)^{-1} C^T B$$

Let $A = \Phi(n)$, $B^{-1} = \lambda \Phi(n-1)$, $C = \mathbf{x}(n)$, $D = 1$, we can write

$$\Phi^{-1}(n) = \lambda^{-1} \Phi^{-1}(n-1) - \lambda^{-1} \left(\frac{\lambda^{-1} \Phi^{-1}(n-1) \mathbf{x}^*(n)}{1 + \lambda^{-1} \mathbf{x}^T(n) \Phi^{-1}(n-1) \mathbf{x}^*} \right) \mathbf{x}^T(n) \Phi^{-1}(n-1)$$

or

$$\mathbf{P}(n) = \lambda^{-1} \mathbf{P}(n-1) - \lambda^{-1} \mathbf{k}(n) \mathbf{x}^T(n) \mathbf{P}(n-1) \tag{A.2}$$

It is now possible to develop the recursive expression for the filter coefficients

$$\begin{aligned}
\mathbf{w}(n) &= \Phi^{-1}(n) \theta(n) \\
&= \Phi^{-1}(n) \left(\lambda \theta(n-1) + d(n) \mathbf{x}^*(n) \right) \\
&= \lambda \left(\lambda^{-1} \Phi^{-1}(n-1) - \lambda^{-1} \mathbf{k}(n) \mathbf{x}^T(n) \Phi^{-1}(n-1) \right) \theta(n-1) + d(n) \Phi^{-1}(n) \mathbf{x}^*(n) \\
&= \Phi^{-1}(n-1) \theta(n-1) - \mathbf{k}(n) \mathbf{x}^T(n) \Phi^{-1}(n-1) \theta(n-1) + d(n) \Phi^{-1}(n) \mathbf{x}^*(n) \\
&= \mathbf{w}(n-1) - \mathbf{k}(n) \mathbf{x}^*(n) \mathbf{w}(n-1) + d(n) \mathbf{k}(n) \\
&= \mathbf{w}(n-1) + \left(d(n) - \mathbf{x}^T(n) \mathbf{w}(n-1) \right) \mathbf{k}(n)
\end{aligned}$$

Equation A.2 is a form of Riccati equation, and we can relate $\mathbf{k}(n) = \mathbf{P}(n) \mathbf{x}^*(n)$ with some algebraic manipulation (Adapted from [230]).

A.2 Linearisation of Non-linear Model

In linearising a nonlinear system, we assume that the variables deviate only slightly from some operating condition. Consider a nonlinear system whose output y is a function of two inputs x_1 and x_2:

$$y = f(x_1, x_2) \tag{A.3}$$

To obtain a linear approximation to this nonlinear system, equation A.3 may be expanded into a Taylor series about the normal operating point $\bar{x}_1, \bar{x}_2$:

$$y = f\left(\bar{x}_1, \bar{x}_2\right) + \left[\frac{\partial f}{\partial x_1}\left(x_1 - \bar{x}_1\right) + \frac{\partial f}{\partial x_2}\left(x_2 - \bar{x}_2\right)\right]$$

$$+ \frac{1}{2!}\left[\frac{\partial^2 f}{\partial x_1^2}\left(x_1 - \bar{x}_1\right)^2 + 2\frac{\partial^2 f}{\partial x_1 \partial x_2}\left(x_1 - \bar{x}_1\right)\left(x_2 - \bar{x}_2\right) + \frac{\partial^2 f}{\partial x_2^2}\left(x_2 - \bar{x}_2\right)^2\right]$$

$$+ \cdots$$

where the partial derivatives are evaluated at $x_1 = \bar{x}_1$, $x_2 = \bar{x}_2$. The higher order terms may be neglected. The linear model of this nonlinear system around the normal operating point can be given by

$$y - \bar{y} = K_1\left(x_1 - \bar{x}_1\right) + K_2\left(x_2 - \bar{x}_2\right)$$

where $\bar{y} = f\left(\bar{x}_1, \bar{x}_2\right)$, $K_1 = \left.\frac{\partial f}{\partial x_1}\right|_{x_1 = \bar{x}_1, x_2 = \bar{x}_2}$ and $K_2 = \left.\frac{\partial f}{\partial x_2}\right|_{x_1 = \bar{x}_1, x_2 = \bar{x}_2}$

The linearisation above is valid in the vicinity of the operating condition. It would not be valid if the operating conditions vary widely (Adapted from [213]).

Example

Linearisation of the "Minimal model for glucose disappearance":

$$\frac{dG}{dt} = -\left(p_1 + X\right)G - p_1 G_b$$

Let $f = \frac{dG}{dt} = -\left(p_1 + X\right)G - p_1 G_b$, then

$$\left.\frac{\partial f}{\partial G}\right|_{G=\bar{G}, X=\bar{X}} = \frac{\partial}{\partial G}\left[-\left(p_1 + X\right)G + p_1 G_b\right] = -\left(p_1 + X\right)|_{X=\bar{X}} = -\left(p_1 + \bar{X}\right)$$

$$\left.\frac{\partial f}{\partial X}\right|_{G=\bar{G}, X=\bar{X}} = \frac{\partial}{\partial X}\left[-p_1 G - XG + p_1 G_b\right] = -G|_{G=\bar{G}} = -\bar{G}$$

$$f - \bar{f} = -\left(p_1 + \bar{X}\right)\left(G - \bar{G}\right) + \left(-\bar{G}\right)\left(X - \bar{X}\right)$$
$$\Rightarrow \quad f = -\left(p_1 + \bar{X}\right)\left(G - \bar{G}\right) + \left(-\bar{G}\right)\left(X - \bar{X}\right) + -\left(p_1 + \bar{X}\right)\bar{G} + p_1 G_b$$
$$= -\left(p_1 + \bar{X}\right)G - \bar{G}\left(X - \bar{X}\right) + p_1 G_b$$

The linearised model is $\dfrac{dG}{dt} = -\left(p_1 + \bar{X}\right)G - \bar{G}\left(X - \bar{X}\right) + p_1 G_b$

B

Model Parameters

B.1 Ackerman's Model

Ackerman's model parameters, as tabulated by Yipintsoi et al [252], is partially reproduced here.

Table B.1. Ackerman's model parameters

Subject	E_G	E_H	m_1	m_2	m_3	m_4
N6	0.09	2.5	0.0351	0.0262	0.0540	0.0262
N4	0.22	6.5	0.0273	0.0271	0.0540	0.0136
N3	0.23	1.3	0.0617	0.0438	0.0590	0.0277
N1	0.41	6.5	0.0251	0.0703	0.0980	0
N2	0.65	11.3	0.0536	0.0858	0.0108	0.0423
N5	0.70	2.6	0.0574	0.1575	0.0729	0
D4a	0.15	8.8	0.0020	0.0014	0.0220	0
D4b	0.08	3.9	0.0009	0.0031	0.0415	0
D2	0.11	4.4	0.0016	0.0143	0.0293	0
D8	0.32	13.5	0.0035	0.0006	0.0418	0
D7a	0.33	364	0.0027	0.0012	0.0260	0
D6	0.33	939	0.0081	0.0857	0.0011	0.0006
D11	0.36	6.9	0.0088	0.0001	0.0338	0
D1	1.10	121	0.0063	0.0011	0.0072	0
D7b	1.60	442	0.0056	0.0011	0.0097	0

- Nn = Normal patient; Dn = diabetic patient.
- E_G and E_H are error values defined as $\sum_{i=1}^{N} \frac{\left(n_i - n_i'\right)^2}{N}$ where n_i = measured data, n_i' = best-fitted data, and N = total number of data points.

F. Chee & T. Fernando: Closed-Loop Control of Blood Glucose, LNCIS 368, pp. 133–137, 2007.
springerlink.com   © Springer-Verlag Berlin Heidelberg 2007

B.2 Minimal Model

Minimal model's parameters, as reported by Bergman et al [197] and Furler et al [176].

Table B.2. Human subject physiological values at time of experiment

Subject	Gender	IBW	G_0	I_0	V_G
Lean		%	mg/dl	μU/ml	dl
1	M	101	94	17	92.6
2	M	100	91	9	125.4
3	M	105	85	15	132.1
4	M	95	99	8	108.4
5	M	100	93	11	127.6
6	M	97	98	4	101.5
7	M	88	97	3	126.5
8	M	98	93	6	118
Obese		%	mg/dl	μU/ml	dl
9	F	153	100	8	92.9
10	F	154	94	15	110.9
11	F	147	96	17	81
12	F	142	110	26	161
13	F	148	99	21	168.4
14	F	130	92	20	135.4
15	F	138	102	81	116.7
16	F	172	109	37	152.8
17	F	206	104	68	195.2
18	F	153	103	16	146
Furler (all)	–	70	81	15	12 L

- IBW = Ideal Body Weight; G_0 = basal glucose value; I_0 = basal insulin value; V_G = glucose distribution space;
- G(0) = BG concentration at the start of clinical experiment;
- "Furler (all)" = values for all three "hypothetical subjects", as presented in Furler et al [176].
  1. "Furler1" = normal patient;
  2. "Furler2" = insulin-resistant patient;
  3. "Furler3" = glucose-resistant patient.

B.3 Cobelli's Model

The parameter values for Cobelli's model considered a normal individual of 70 kg body weight, with plasma blood glucose $y_1 = 91.5$ mg/dl, plasma insulin concentration $y_2=11$ μU/ml, and blood glucagon concentration $y_5=75$ pg/ml. $u_{1p}(0)=4.9$ U and $u_{2p}(0)=0.49$ U.

Table B.3. Minimal Model parameters

Subject	G(0)	p_1	p_2	$p_3(\times 10^6)$	I(0)	$\gamma(\times 10^3)$	h	n
Lean	mg/dl	min^{-1}	min^{-1}	min^{-2}/(μU/ml)	μU/ml	μU/ml min^2	mg/dl	min^{-1}
1	298	0.0296	0.0186	6.51	333	5.36	90.9	0.23
2	276	0.0192	0.0262	14.7	69	1.40	100.0	0.18
3	250	0.0374	0.0478	8.73	33	2.93	87.5	0.30
4	337	0.0363	0.0081	4.01	192	2.40	93	0.23
5	329	0.0464	0.0038	3.61	73	1.69	119	0.13
6	296	0.0136	0.0341	17.3	50	0.89	90.9	0.22
7	248	0.0151	0.0313	9.7	15	0.69	82.6	0.22
8	271	0.0217	0.0292	19.1	14	0.28	87.3	0.11
Obese								
9	297	0.0400	0.0042	2.56	209	3.72	154	0.22
10	242	0.0323	0.0034	2.67	104	3.46	145	0.19
11	348	0.0466	0.0676	16.1	264	7.47	186	0.09
12	256	0.0180	0.0108	2.29	99	3.42	153	0.13
13	217	0.0113	0.0034	0.97	225	3.30	122	0.14
14	224	0.0113	0.0240	4.89	185	1.36	123	0.11
15	258	0.0246	0.0069	0.55	337	2.70	175	0.13
16	267	0.0100	0.0166	2.02	248	3.40	108	0.13
17	242	0.0071	0.0125	1.58	20	6.11	143	0.13
18	254	0.0093	0.0200	7.78	16	1.19	132	0.13
Furler1	–	0.028	0.0250	13	–	–	–	0.09
Furler2	–	0.028	0.0250	5	–	–	–	0.09
Furler3	–	0	0.0250	13	–	–	–	0.09

Table B.4. Parameter values for pathological states and other conditions

Overt diabetes	Latent diabetes	Obesity
F_2=0.037 constant	F_2=0.037 constant	H_4=0.0012 constant
H_4=0.0012 constant	H_4=0.0012 constant	$b_{42} = 7 \times 10^{-3}$
$b_{42} = 7 \times 10^{-3}$	$b_{42} = 7 \times 10^{-3}$	c_{42}=40.47
c_{42}=-40.47	c_{42}=-40.47	m_{02}=0.13
b_w=4.5$\times 10^{-3}$	b_w=4.5$\times 10^{-3}$	b_6=0.5
b_6=5$\times 10^{-3}$	b_6=1$\times 10^{-2}$	c_6=-3.64
c_w=-306.25	c_w=-306.25	
c_6=-363.55	c_6=-176.68	

B.4 Hovorka's Model

Hovorka's model constants and parameters (primarily from reference [211])

Table B.5. Hovorka's model constants

Symbol	Quantity	Value
k_{12}	Transfer rate	$0.066\,\mathrm{min}^{-1}$
k_{a1}	Deactivation rate	$0.006\,\mathrm{min}^{-1}$
k_{a2}	Deactivation rate	$0.06\,\mathrm{min}^{-1}$
k_{a3}	Deactivation rate	$0.03\,\mathrm{min}^{-1}$
k_e	Insulin elimination from plasma	$0.138\,\mathrm{min}^{-1}$
V_I	Insulin distribution volume	$0.12\,\mathrm{L\,kg}^{-1}$
V_G	Glucose distribution volume	$0.16\,\mathrm{L\,kg}^{-1}$
A_G	Carbohydrate bio-availability	0.8 (unitless)
$t_{max,G}$	Time-to-maximum of carbohydrate absorption	40 min

Table B.6. Hovorka's model parameters for the purpose of Bayesian parameter estimation

Symbol	Quantity	Mean Value
S_{IT}^f	Insulin sensitivity of distribution/transport	$51.2 \times 10^{-4}\,\mathrm{min}^{-1}$ per mU/l
S_{ID}^f	Insulin sensitivity of disposal	$8.2 \times 10^{-4}\,\mathrm{min}^{-1}$ per mU/l
S_{IE}^f	Insulin sensitivity of EGP	520×10^{-4} per mU/l
EGP_0	EGP extrapolated to zero insulin concentration	$0.0161\,\mathrm{mmol \cdot kg}^{-1} \cdot \mathrm{min}^{-1}$
F_{01}	Insulin-independent glucose flux	$0.0097\,\mathrm{mmol \cdot kg}^{-1} \cdot \mathrm{min}^{-1}$
$t_{max,I}$	Time-to-maximum of absorption of subcutaneously injected short-acting insulin	55 min

B.5 Sorensen's Model

The nomenclature used in Sorensen's model parameter, as according to [209]:

$i(t)$ = insulin rate entering patient [mU/min];

G = glucose concentration [mg/dl];

I = insulin concentration [mU/dl];

χ = glucagon concentration [pg/ml].

V = volume [dl or L];

Q = vascular blood water flow rate [dl/min or L/min];

Γ = metabolic source or sink rate [mg/min];

T = trans-capillary diffusion rate [min];

t = time [min];

- The first subscript represents physiological compartment, i.e., B (brain), G (gut), K (kidney), L (Liver), H (heart), PV (Periphery), A (hepatic artery)
- Second subscript are defined as: I = interstitial fluid space; V = vascular blood water space;

- Metabolic rate subscripts:
  PGU = peripheral glucose uptake;
  RBCU = red blood cell glucose uptake;
  PIR = peripheral insulin release;
  LIC = liver insulin clearance;
  KIC = kidney insulin clearance;
  PIC = peripheral insulin clearance;
  KGE = kidney glucose excretion;
  HGU = hepatic glucose uptake;
  HGP = hepatic glucose production;
  GGU = gut glucose utilisation;
  BGU = brain glucose uptake;
  $P_\chi C$ = plasma glucagon clearance;
  $M_\chi C$ = metabolic glucagon clearance;
  $P_\chi R$ = pancreatic glucagon release;
- First superscript are defined as: G = glucose; I = insulin; χ = glucagon; B = basal value; N = normalised value (divided by basal value)

Table B.7. Values for metabolic sources/sinks

$\Gamma_{BGU} = 70$ mg/min
$\Gamma_{RBGU} = 10$ mg/min
$\Gamma_{GGU} = 20$ mg/min
$\Gamma_{PGU}^{B} = 35$ mg/min
$\Gamma_{HGP}^{B} = 155$ mg/min
$\Gamma_{HGU}^{B} = 20$ mg/min
$G_{PI}^{N} = \frac{G_{PI}}{\text{basal value}}$ (e.g. basal value = 86.81)
$I_{PI}^{N} = \frac{I_{PI}}{\text{basal value}}$ (e.g. basal value = 5.304)
$G_{L}^{N} = \frac{G_{L}}{\text{basal value}}$ (e.g. basal value = 101)
$I_{L}^{N} = \frac{I_{L}}{\text{basal value}}$ (e.g. basal value = 21.43)

Table B.8. Sorensen's model parameter values for the diabetic patient

$V_{BV}^{G} = 3.5$ dL	$Q_B^G = 5.9$ dL/min	$V_B^I = 0.26$ L	$Q_B^I = 0.45$ L/min	$\tau_I = 25$ min
$V_H^G = 13.8$ dL	$Q_H^G = 43.7$ dL/min	$V_H^I = 0.99$ dL	$Q_H^I = 3.12$ L/min	$\tau_\chi = 65$ min
$V_G^G = 11.2$ dL	$Q_G^G = 10.1$ dL/min	$V_G^I = 0.94$ L	$Q_G^I = 0.72$ L/min	$T_B = 2.1$ min
$V_L^G = 25.1$ dL	$Q_L^G = 12.6$ dL/min	$V_L^I = 1.14$ L	$Q_L^I = 0.9$ L/min	$T_P^G = 5.0$ min
$V_K^G = 6.6$ dL	$Q_K^G = 10.1$ dL/min	$V_K^I = 0.51$ L	$Q_K^I = 0.72$ L/min	$T_P^I = 20$ min
$V_{PV}^G = 10.4$ dL	$Q_P^G = 15.1$ dL/min	$V_{PV}^I = 0.74$ L	$Q_P^I = 1.05$ L/min	$V_{BI} = 4.5$ dL
	$Q_A^G = 2.5$ dL/min		$Q_A^I = 0.18$ L/min	$V_{PI} = 63$ dL

References

1. J. Desemore, J.P. Mordes, and A.A. Rossini. Hypoglycemia. In J.M. Rippe, R.S. Irwin, J.S. Alpert, and M.P. Fink, editors, *Intensive Care Medicine 2nd Ed.*, chapter 99, pages 1000–1009. Little, Brown and Company, 1991.
2. I.B. Hirsch and J.B. McGill. Role of insulin in management of surgical patients with diabetes melliltus. *Diabetes Care*, 13(9):980–991, September 1990.
3. W.F. Ganong. *Review of Medical Physiology*. Appleton & Lange, Stanford, Connecticut, 1997.
4. A.C. Guyton and J.E. Hall. *Textbook of Medical Physiology*. W.B. Saunders Company, 9 edition, 1996.
5. I.A. Macfarlane. *The Millenia Before Insulin*, volume 1, chapter 1, pages 3–9. Blackwell Scientific Publications, 1991.
6. M. Bliss. *The Discovery of Insulin*, volume 1, chapter 2, pages 10–14. Blackwell Scientific Publications, 1991.
7. B.A. Mizock. Alterations in carbohydrate metabolism during stress: A review of the literature. *The American Journal of Medicine*, 98:75–84, January 1995.
8. N.B. Watts, S.S.P. Gebhart, R.V. Clark, and L.S. Phillips. Postoperative management of diabetes mellitus: Steady-state glucose control with bedside algorithm for insulin adjustment. *Diabetes Care*, 10(6):722–728, Nov - Dec 1987.
9. G. Van den Berghe, P. Wouters, F. Weekers, C. Verwaest, F. Bruyninckx, M. Schetz, D. Vlasselaers, P. Ferdinande, P. Lauwers, and R. Bouillon. Intensive insulin therapy in critically ill patients. *N Engl J Med*, 345(19):1359–1367, 2001.
10. A.M. Albisser and B.S. Leibel. Artificial beta cell, 1981. US Patent 4,245,643.
11. The Diabetes Control and Complications Trial Research Group. The effect of intensive treatment of diabetes on the development and progression of long-term complications in insulin-dependent diabetes mellitus. *N Engl J Med*, 329(14):977–986, 30 September 1993.
12. UK Prospective Diabetes Study (UKPDS) Group. Intensive blood-glucose control with sulphonylureas or insulin compared with conventional treatment and risk of complications in patients with type 2 diabetes (UKPDS 33). *The Lancet*, 352:837–853, 12 September 1998.
13. Y. Ohkubo, H. Kishikawa, E. Araki, T. Miyata, S. Isami, S. Motoyoshi, Y. Kojima, N. Furuyoshi, and M. Shichiri. Intensive insulin therapy prevents the progression of diabetic microvascular complications in Japanese patients with non-insulin-dependent diabetes mellitus: a randomised prospective 6-year study. *Diabetes Research and Clinical practice*, 28:103–117, May 1995.

14. S.H. Golden, C. Peart-Vigilance, W.H. L. Kao, and F.L. Brancati. Perioperative glycemic control and the risk of infectious complications in a cohort of adults with diabetes. *Diabetes Care*, 22(9):1408–1414, September 1999.

15. L.C. Gatewood, E. Ackerman, J.W. Rosevear, and G.D. Molnar. Simulation studies of blood-glucose regulation: Effect of intestinal glucose absorption. *Computers and Biomedical Research*, 2:15–27, 1968.

16. A.H. Clemens, P.H. Chang, and R.W. Myers. The development of Biostator, a glucose-controlled insulin infusion systems (GCIIS). *Horm Metab Res Suppl*, 7:23–33, March 1977.

17. A.M. Albisser, B.S. Leibel, T.G. Ewart, Z. Davidovac, C.K. Botz, and W. Zingg. Clinical control of diabetes by the artificial pancreas. *Diabetes*, 23(5):397–404, May 1974.

18. D.J. Chisolm, E.W. Kraegen, D.J. Bell, and D.R. Chipps. A semi-closed loop computer-assisted insulin infusion system: Hospital use for control of diabetes in patients. *Med J Aust*, 141:784–788, 1984.

19. L. Heinemann and F. J. Ampudia-Blasco. Glucose clamps with the Biostator: A critical reappraisal. *Horm Metab Res*, 26:579–583, 1994.

20. U. Fischer, K. Rebrin, T.v. Woedtke, and P. Abel. Clinical usefulness of the glucose concentration in the subcutaneous tissue - properties and pitfalls of electrochemical biosensors. *Horm Metab Res*, 26:515–522, 1994.

21. U. Fischer. Continuous in vivo monitoring in diabetes: the subcutaneous glucose concentration. *Acta Anaesthesiol Scand Suppl*, 39(104):21–29, 1995.

22. E. F. Pfeiffer. Review. on the way to the automated (blood) glucose regulation in diabetes: the dark past, the gray present and the rosy future. *Diabetologia*, 30:51–65, 1987.

23. A.M. Albisser. Insulin delivery systems: Do they need a glucose sensor? *Diabetes Care*, 5(3):166–173, May-June 1982.

24. B.H. Ginsberg. An overview of minimally invasive technologies. *Clin. Chem.*, 38(9):1596–1600, 1992.

25. K. Rebrin, G. M. Steil, William P. Van Antwerp, and John J. Mastrototaro. Subcutaneous glucose predicts plasma glucose independent of insulin: Implications for continuous monitoring. *Am J Physiol Endocrinol. Metab. 40*, 277:E561–E571, 1999.

26. C. Meyerhoff, E. Bischf, F. Sternberg, H. Zier, and E.F. Pfeiffer. On line continuous monitoring of subcutaneous tissue glucose in men by combining portable glucosensor with microdialysis. *Diabetologia*, 35:1087–1092, 1992.

27. A. Heller. Implanted electrochemical glucose sensor for the management of diabetes. *Annu. Rev. Biomed. Eng.*, 1:153–175, 1999.

28. B.H. Ginsberg. The role of technology in diabetes therapy. *Diabetes Care*, 17 Suppl. 1:50–55, June 1994.

29. Editorial. Problems associated with subcutaneously implanted glucose sensors. *Diabetes Care*, 22(2):143–148, February 1999.

30. M. Shichiri, R. Kawamori, Y. Goriya, Y. Yamasaki, M. Nomura, N. Hakui, and H. Abe. Glycaemic control in pancreatectomized dogs with a wearable artificial endocrine pancreas. *Diabetologia*, 24:179–184, 1983.

31. M. Shichiri, R. Kawamori, Y. Goriya, Y. Yamasaki, M. Nomura, N. Hakui, and H. Abe. Closed-loop glycaemic control with a wearable artificial endocrine pancreas - variations in daily insulin requirements to glycemic response. *Diabetes*, 33:1200–1202, December 1984.

32. Tran Minh Canh. *Biosensors*, volume 1 of *Sensor Physics and Technology Series*, chapter 2, pages 7–19. Chapman & Hall and Masson, 2-6 Boundary Row, London SE1 8HN, 1993.

33. Gbor Harsnyi. *Sensors in Biomedical Applications, - Fundamentals, Technology & Applications*. Technomic Publishing Company, Inc., Pennsylvania USA, 2000.

34. MiniMed. MiniMed glucose sensor - instructions for use (REF MMT-7002). Northridge, CA, May 2001.

35. V. Poitout, D. Moatti-Sirat, G. Reach, Y. Zhang, G.S. Wilson, F. Lemonnier, and J. C. Klein. A glucose monitoring system for on-line estimation in man of blood glucose concentration using a miniaturized glucose sensor implanted in the subcutaneous tissue and a wearable control unit. *Diabetologia*, 36:658–663, 1993.

36. J.S. Schultz, S. Mansouri, and I.J. Goldstein. Affinity sensor: A new technique for developing implantable sensors for glucose and other metabolites. *Diabetes Care*, 5(3):245–253, 1982.

37. J.I. Peterson and G.G. Vurek. Fiber-optic sensors for biomedical applications. *Science*, 224(4645):123–127, 1984.

38. R.J. McNichols and G.L. Coté. Optical glucose sensing in biological fluids: an overview. *Journal of Biomedical Optics*, 5(1):5–16, January 2000.

39. L.J. McCartney, J.C. Pickup, O.J. Rolinski, and D.J.S. Birch. Near-infrared fluorescence lifetime assay for serum glucose based on Allophycocyanin-Labeled Concanavalin A. *Analytical Biochemistry*, 292:216–221, 2001.

40. J.C. Pickup, F. Hussain, N.D. Evans, O.J. Rolinski, and D.J.S. Birch. Fluorescence-based glucose sensors. *Biosensors and Bioelectronics*, 20:2555–2565, 2005.

41. J. Pickup, L. McCartney, O. Rolinski, and D. Birch. In vivo glucose sensing for diabetes management: Progress towards non-invasive monitoring. *BMJ*, 319:1–4, 13 November 1999.

42. R. Ballerstadt and J.S. Schultz. A fluorescence affinity hollow fiber sensor for continuous transdermal glucose monitoring. *Analytical Chemistry*, 72:4185–4192, 2000.

43. J.S. Kristensen, S. Aasmul, J.K. Nielsen, and J.S. Christiansen. Transcutaneous fluorescence lifetime based continuous glucose reading for long term interrogation, 2005. Poster at 65th annual American Diabetes Association meeting 2005, Download from http://www.precisense.com/graphics/Synkron-Library/Products/ presicense_handout_higress.pdf.

44. W.F. March, K. Ochsner, and J. Horna. Intraocular lens glucose sensor. *Diabetes Technology & Therapeutics*, 2(1):27–30, 2000.

45. L. Li and D.R. Walt. Dual-analyte fiber-optic sensor for the simultaneous and continuous measurement of glucose and oxygen. *Analytical Chemistry*, 67:3746–3752, 1995.

46. H. Gunasingham, Chin-Huat Tan, and J.K.L. Seow. Fiber-optic glucose sensor with electrochemical generation of indicator reagent. *Analytical Chemistry*, 62:755–759, 1990.

47. M.S. Abdel-Latif and G.G. Guilbault. Fiber-optic sensor for the determination of glucose using micellar enhanced chemiluminescence of the peroxyoxalate reaction. *Analytical Chemistry*, 60:2671–2674, 1988.

48. N. Wisniewski, F. Moussy, and W.M. Reichert. Characterization of implantable biosensor membrane biofouling. *Fresenius J Anal Chem*, 366:611–621, 2000.

49. M.J. McShane. Potential for glucose monitoring with nanoengineered fluorescent biosensors. *Diabetes Technology & Therapeutics*, 4(4):533–538, 2002.

50. E.F. Pfeiffer. The "Ulm Zucker Uhr System" and its consequences. *Horm Metab Res*, 26:510–514, 1994.

51. J.A. Tamada, M. Lesho, and M.J. Tierney. Keeping watch on glucose - New monitors help fight the long-term complications of diabetes. *IEEE Spectrum*, 39(4):52–57, 2002.

52. S. Mansouri and J.S. Schultz. A miniature optical glucose sensor based on affinity binding. *Biotechnology*, 2:885–890, October 1984.

53. J.C. Pickup. In vivo glucose monitoring: Sense and sensorbility. *Diabetes Care*, 16(2):535–539, 1993.

54. D.C. Klonoff. Noninvasive blood glucose monitoring. *Diabetes Care*, 20(3):433–437, March 1997.

55. J.A. Tamada, S. Garg, L. Jovanovic, K.R. Pitzer, S. Fermi, R.O. Potts, and the Cygnus Research Team. Noninvasive glucose monitoring - Comprehensive clinical results. *JAMA*, 282(19):1839–1844, November 1999.

56. M.J. Tierney, Y. Jayalakshmi, N.A. Parris, M.P. Reidy, C. Uhegbu, and P. Vijayakumar. Design of a biosensor for continual, transdermal glucose monitoring. *Clinical Chemistry*, 45(9):1681–1683, 1999.

57. S.K. Garg, R.O. Potts, N.R. Ackerman, S.J. Fermi, J.A. Tamada, and H.P Chase. Correlation of Fingerstick blood glucose measurements with GlucoWatch biographer glucose results in young subjects with type 1 diabetes. *Diabetes Care*, 22(10):1708–1712, October 1999.

58. J. Kost. Ultrasound-assisted insulin delivery and noninvasive glucose sensing. *Diabetes Technology & Therapeutics*, 4(4):489–497, 2002.

59. S. Mitragotri, M. Coleman, J. Kost, and R. Langer. Transdermal extraction of analytes using low-frequency ultrasound. *Pharmaceutical Research*, 17(4):466–470, 2000.

60. J. Kost, S. Mitragotri, R.A. Gabbay, M. Pishko, and R. Langer. Transdermal monitoring of glucose and other analytes using ultrasound. *Nature Medicine*, 6(3):347–350, March 2000.

61. S. Gebhart, M. Faupel, R. Fowler, C.Kapsner, D. Lincoln, V. McGee, J. Pasqua, L. Steed, M. Wangsness, F. Xu, and M. Vanstory. Glucose sensing in transdermal body fluid collected under continuous vacuum pressure via micropores in the stratum corneum. *Diabetes Technology & Therapeutics*, 5(2):159–166, 2003.

62. J.A. Eppstein, M.R. Hatch, and D. Yang. Microporation of human skin for drug delivery and monitoring applications, 2000. United States Patent US 6,142,939.

63. L. Heinemann and G. Schmelzeisen-Redeker on behalf of the Non-invasive task force *NITF*. Non-invasive continuous glucose monitoring in type I diabetic patients with optical glucose sensors. *Diabetologia*, 41:848–854, 1998.

64. J.W. Hall and A. Pollard. Near-infrared spectrophotometry: A new dimension in clinical chemistry. *Clinical Chemistry*, 38(9):1623–1631, 1992.

65. John A Dean. *Dean's Analytical Chemistry Handbook*. McGraw-Hill, New York, 2nd edition, 2004.

66. Y. Mendelson, A.C. Clermont, R.A. Peura, and Been-Chyuan Lin. Blood glucose measurement by multiple attenuated total reflection and infrared absorption spectroscopy. *IEEE Trans. Biomed. Eng.*, 37(5):458–465, May 1990.

67. D.A. Skoog and J.J. Leary. *Principles of Instrumental Analysis*. Saunders College Publishing, Florida, USA, 4th edition, 1992.

68. K. Danzer, Ch. Fischbacher, K.U. Jagemann, and K.J. Reicheit. Near-infrared diffuse reflection spectroscopy for non-invasive blood glucose monitoring. *IEEE-LEOS Newsletter (Special issue)*, pages 9–11, April 1998.

69. Yoen-Joo Kim and G. Yoon. Prediction of glucose in whole blood by near-infrared spectroscopy: Influence of wavelength region, preprocessing, and hemoglobin concentration. *Journal of Biomedical Optics*, 11(4):041128, July/August 2006.

70. J.J. Burmeister and M.A. Arnold. Spectroscopic considerations for noninvasive blood glucose measurements with near infrared spectroscopy. *IEEE-LEOS Newsletter (Special issue)*, pages 6–9, April 1998.

71. G. Yoon, A.K. Amerov, K.J. Jeon, J.B. Kim, and Y-J Kim. Optical measurement of glucose levels in scattering media. volume 20, pages 1897–1899. IEEE, 1998.

72. R. Abbink and C. Gardner. Getting under the skin. *spie's oemagazine*, pages 18–20, September 2003.

73. N. Kaiser. Laser absorption spectroscopy with an ATR prism - noninvasive in vivo determination of glucose. *Horm Metab Res Suppl*, 8:30–33, 1979.

74. P. Zheng, C.E. Kramer, C.W. Barnes, J.R. Braig, and B.B. Sterling. Noninvasive glucose determination by oscillating thermal gradient spectrometry. *Diabetes Technology & Therapeutics*, 2(1):17–26, 2000.

75. G.B. Christison and H.A. MacKenzie. Laser photoacoustic determination of physiological glucose concentrations in human whole blood. *Medical & Biological Engineering & Computing*, 31(3):284–290, 1993.

76. H.A. MacKenzie, H.S. Ashton, S. Spiers, Y. Shen, S.S Freeborn, J. Hannigan, J. Lindberg, and P. Rae. Advances in photoacoustic noninvasive glucose testing. *Clinical Chemistry*, 45(9):1587–1595, 1999.

77. F.M. Ham, G.M. Cohen, K. Patel, and B.R. Gooch. Multivariate determination of glucose using NIR spectra of human blood serum. In *Proceedings of the 16th Annual International Conference of the IEEE on Engineering Advances: New Opportunities for Biomedical Engineers*, volume 2, pages 818–819, Baltimore, MD, USA, 03 November-06 November 1994. IEEE, IEEE Engineering in Medicine and Biology Society, 1994.

78. Ch. Fischbacher, K.-U. Jagemann, K. Danzer, U.A. Müller, L. Papenkordt, and J. Schüler. Enhancing calibration models for non-invasive near-infrared spectroscopical blood glucose determination. *Fresenius J Anal Chem*, 359:78–82, 1997.

79. M.R. Robinson, R.P. Eaton, D.M. Haaland, G.W. Koepp, and E.V. Thomas. Noninvasive glucose monitoring in diabetic patients: A preliminary evaluation. *Clinical Chemistry*, 38(9):1618–1622, 1992.

80. G.W. Small, M.A. Arnold, and L.A. Marquardt. Strategies for coupling digital filtering with partial least-squares regression: Application to the determination of glucose in plasma by Fourier Transform Near Infrared spectroscopy. *Analytical Chemistry*, 65:3279–3289, 1993

81. H.M. Heise and R. Marbach. Multivariate determination of glucose in whole blood by attenuated total reflection infrared spectroscopy. *Analytical Chemistry*, 61:2009–2015, 1989.

82. J. Tenhunen, H. Kopola, and R. Myllylä. Non-invasive glucose measurement based on selective near infrared absorption requirements on instrumentation and spectral range. *Measurement*, 24:173–177, 1998.

83. PerkinElmer Inc. FT-IR spectroscopy: Attenuated total reflectance (ATR) - Technical Note, 2007. [Online] Available at http://las.perkinelmer.com/content/TechnicalInfo/TCH_FTIRATR.pdf.

84. H.M. Heise. Non-invasive monitoring of metabolites using near infrared spectroscopy: State of the art. *HormMetabRes*, 28:527–534, 1996.

85. G.M. Oosta, Tayy-Wen Jeng, J.M. Lindberg, M.L. McGlashen, and J.L. Pezzaniti. Multiplex sensor and method of use, 2003. United States Patent US 6,567,678 B1.

86. C.D. Malchoff, K. Shoukri, J.I. Landau, and J.M. Buchert. A novel noninvasive blood glucose monitor. *Diabetes Care*, 25(12):2268–2275, December 2002.
87. D.C. Klonoff and J. Braig. Mid-infrared spectroscopy for noninvasive blood glucose monitoring. *IEEE-LEOS Newsletter (Special issue)*, pages 13–14, April 1998.
88. J.R. Braig, C.E. Kramer, B.B. Sterling, D.S. Goldberger, P. Zheng, A.M. Shulenberger, R. Trebino, R.A. King, and C.W. Barnes. Method for determining analyte concentration using periodic temperature modulation and phase detection, 2006. United States Patent US 7,006,857 B2.
89. L. Heinemann, U. Krämer, Hans-Martin Klötzer, M. Hein, D. Volz, M. Hermann, T. Heise, and K. Rave the Non-invasive task force *NITF*. Non-invasive glucose measurement by monitoring of scattering coefficient during oral glucose tolerance tests. *Diabetes Technology & Therapeutics*, 2(2):211–220, 2000.
90. S.L. Jacques and S.A. Prahl. ECE532 Biomedical Optics, Oregon Graduate Institute, 1998. [Online] Available at http://www.phas.ubc.ca/~mackay/phys404/documents/BiomedicalOptics/ece532/class3/mie.html.
91. J.T. Bruulsema, J.E. Hayward, T.J. Farrell, and M.S. Patterson. Correlation between blood glucose concentration in diabetics and noninvasively measured tissue optical scattering coefficient. *Optics Letters*, 22(3):190–192, February 1997.
92. M. Kohl, M. Essenpreis, and M. Cope. The influence of glucose concentration upon the transport of light in tissue-simulating phantoms. *Phy. Med. Biol*, 40:1267–1287, 1995.
93. J.S. Maier, S.A. Walker, S. Fantini, M.A. Franceschini, and E. Gratton. Possible correlation between blood glucose concentration and the reduced scattering coefficient of tissues in the near infrared. *Optics Letters*, 19(24):2062–2064, December 1994.
94. O.S. Khalil. Spectroscopic and clinical aspects of noninvasive glucose measurements. *Clin. Chem.*, 45(2):165–177, 1999.
95. M. Kinnunen, R. Myllylä, T. Jokela, and S. Vainio. In vitro studies toward noninvasive glucose monitoring with optical coherence tomography. *Applied Optics*, 45(10):2251–2260, 1 April 2006.
96. H.M. Heise. Diffuse reflectance near-infrared spectrometry for non-invasive blood glucose monitoring. *IEEE-LEOS Newsletter (Special issue)*, pages 11–13, April 1998.
97. A. Kienle, L. Lilge, M.S. Patterson, R. Hibst, and R. Steiner. Spatially resolved absolute diffuse reflectance measurements for noninvasive determination of the optical scattering and absorption coefficients of biological tissue. *Applied Optics*, 35(13):2304–2314, 1 May 1996.
98. K.V. Larin, M.S. Eledrisi, M. Motamedi, and R.O. Esenaliev. Noninvasive blood glucose monitoring with optical coherence tomography. *Diabetes Care*, 25(12):2263–2267, December 2002.
99. E.M.C. Hillman. *Optical Tomography*, volume 4, pages 2203–2235. Wiley-VCH Verlag GmBH & Co. KGaA, 2003.
100. D. Huang, E.A. Swanson, C.P. Lin, J.S. Schuman, W.G. Stinson, W. Chang, M.R. Hee, T. Flotte, K. Gregory, C.A. Puliafito, and J.G. Fujimoto. Optical coherence tomography. *Science*, 254(5035):1178–1181, 22 November 1991.
101. Ilya Fine. Non-invasive method and system of optical measurements for determining the concentration of a substance in blood, 2002. United States Patent US 6,400,972 B1.
102. I. Fine, B. Fikhte, and L.D. Shvartsman. Occlusion spectroscopy as a new paradigm for non-invasive blood measurements. *Proceedings of SPIE*, 4263:122–130, 2001.

103. O. Cohen, I. Fine, E. Monashkin, and A. Karasik. Glucose correlation with light scattering patterns - novel method for non-invasive glucose measurements. *Diabetes Technology & Therapeutics*, 5(1):11–17, 2003.

104. L.D. Shvartsman and I. Fine. Light scattering changes caused by RBC aggregation: Physical basis for new approach to non-invasive blood count. *Proceedings of SPIE*, 4263:131–142, 2001.

105. K. Yamakoshi and Y. Yamakoshi. Pulse glucometry: a new approach for noninvasive blood glucose measurement using instantaneous differential near-infrared spectrophotometry. *Journal of Biomedical Optics*, 11(5):054028, September/October 2006.

106. (NJ, USA) Princeton Instruments Inc and (MA, USA) Acton Research Corp. Raman spectroscopy basics, 2006. [Online] Available at http://architect.wwwcomm.com/Uploads/Princeton/Documents/ Library/App_Wht_Papers/Raman_Basics.pdf.

107. USA Kaiser Optical Systems Inc, MI. Raman Technical Resources - Raman Tutorial, 2006. [Online] Available at http://www.kosi.com/raman/resources/tutorial/index.html.

108. Tae-Wong Koo, A.J. Berger, I. Itzkan, G. Horowitz, and M.S. Feld. Measurement of glucose in human blood serum using Raman spectroscopy. *IEEE-LEOS Newsletter (Special issue)*, pages 18–19, April 1998.

109. R.J. McNichols, B.D. Cameron, and G.L. Coté. Development of a non-invasive polarimetric glucose sensor. *IEEE-LEOS Newsletter (Special issue)*, pages 30–31, April 1998.

110. J. Lambert, M. Storrie-Lombardi, and M. Borchert. Measurement of physiologic glucose levels using Raman spectroscopy in a rabbit aqueous humor model. *IEEE-LEOS Newsletter (Special issue)*, pages 19–22, April 1998.

111. A.J. Berger, Tae-Woong Koo, I. Itzkan, G. Horowitz, and M.S. Feld. Multi-component blood analysis by near-infrared Raman spectroscopy. *Applied Optics*, 38(13):2916–2926, 1 May 1999.

112. L.D. Barron. *Molecular Light Scattering and Optical Activity*. Cambridge University Press, Cambridge CB2 1RP, 1982.

113. Gordan G. Hammes. *Spectroscopy for the biological sciences*. John Wiley & Sons Inc, New Jersey, USA, 1sh edition, 2005.

114. B.D. Cameron and G.L. Coté. Noninvasive glucose sensing utilizing a digital closed-loop polarimetric approach. *IEEE Trans. Biomed. Eng.*, 44(12):1221–1227, December 1997.

115. G.L. Cote, M.D. Fox, and R.B. Northrop. Optical polarimetric sensor for blood glucose measurement. *Proceedings of the 1990 Sixteenth Annual Northeast Bioengineering Conference, 1990*, pages 101–102, 26-27 March 1990.

116. G.L. Cote, M.D. Fox, and R.B. Northrop. Laser polarimetric sensor for glucose monitoring. *Proceedings of the Twelfth Annual International Conference of the IEEE Engineering in Medicine and Biology Society*, pages 476–477, 1-4 November 1990.

117. H. Anumula, A. Nezhuvingal, Y. Li, , and B.D. Cameron. Development of a noninvasive corneal birefringence compensated glucose sensing polarimeter. *SPIE–The International Society for Optical Engineering*, 4958:303–312, February 2003.

118. G.L. Cote, M.D. Fox, and R.B. Northrop. Noninvasive optical polarimetric glucose sensing using a true phase measurement technique. *IEEE Trans. Biomed. Eng.*, 39(7):752–756, July 1992.

119. G.L. Cote, M.D. Fox, and R.B. Northrop. Optical glucose sensing apparatus and method, 1993. United States Patent US 5,209,231.

120. A.F. Browne, T.R. Nelson, and R.B. Northrop. Microdegree polarimetric measurement of glucose concentrations for biotechnology applications. *Proceedings of the IEEE 1997 23rd Northeast Bioengineering Conference, 1997*, pages 9–10, 21-22 May 1997.

121. C. Chou, Chien-Yuan Han, Wen-Chuan Kuo, Yen-Chuen Huang, Ching-Mei Feng, and Jenn-Chyang Shyu. Noninvasive glucose monitoring in vivo with an optical heterodyne polarimeter. *Applied Optics*, 37(16):3553–3557, 1998.

122. C. Chou and Po-Kang Lin. Noninvasive glucose monitoring with optical heterodyne technique. *Diabetes Technology & Therapeutics*, 2(1):45–47, 2000.

123. S. Jang and M.D. Fox. Optical sensor using the magnetic optical rotatory effect of glucose. *IEEE-LEOS Newsletter (Special issue)*, pages 28–30, April 1998.

124. R.W. Waynant and V.M. Chenault. Overview of non-invasive fluid glucose measurement using optical techniques to maintain glucose control in diabetes mellitus. *IEEE-LEOS Newsletter (Special issue)*, pages 3–6, April 1998.

125. Andrew C. Tam. Applications of photoacoustic sensing techniques. *Reviews of Modern Physics*, 58(2):381–431, April 1986.

126. R. Nagar, B. Pesach, and U. Ben-Ami. Photoacoustic assay and imaging system, 2005. United States Patent US 6,846,288 B2.

127. C. Ruff, S. Lohmann, H. Lubatschowski, and W. Ertner. Photoacoustic determination of optical parameters of biological tissue. pages 325–325. IEEE, 8-13 September 1996.

128. Zuomin Zhao. *Pulsed photoacoustic techniques and glucose determination in human blood and tissue.* ISBN 951-42-6689-7. Oulu University Press, Oulu, 2002. [Online] Available from http://herkules.oulu.fi/isbn9514266900/.

129. Z. Zhao and R.A. Myllylä. Photoacoustic blood glucose and tissue measurements based on optical scattering effect. *Proceedings of the SPIE*, 4707:153–157, July 2002.

130. K.M. Quan, E.M. Johnston, and H.A. MacKenzie. Noninvasive glucose monitoring using nearinfrared photoacoustic spectroscopy. pages 264–264. IEEE, 28 August-2 September 1994.

131. G. Spanner and R. Nießner. Noninvasive determination of blood constituents using an array of modulated laser diodes and a photoacoustic sensor head. *Fresenius J Anal Chem*, 355:327–328, 1996.

132. G. Spanner and R. Nießner. New concept for the non-invasive determination of physiological glucose concentrations using modulated laser diodes. *Fresenius J Anal Chem*, 354:306–310, 1996.

133. R. Weiss, Y. Yegorchikov, A. Shusterman, and I. Raz. Noninvasive continuous glucose monitoring using photoacoustic technology - Results from the first 62 subjects. *Diabetes Technology & Therapeutics*, 9(1):68–74, 2007.

134. S.A. Asher and J.H. Holtz. Glucose sensing intelligent polymerized crystalline colloidal arrays. *IEEE-LEOS Newsletter (Special issue)*, pages 32–34, April 1998.

135. A.J. Marshall, D.S. Young, J. Blyth, S. Kabilan, and C.R. Lowe. Metabolite-sensitive holographic biosensors. *Analytical Chemistry*, 76:1518–1523, 2004.

136. S. Kabilan, A.J. Marshall an F.K. Sartain, M. Lee, A. Hussain, X. Yang, J. Blyth, N. Karangu, K. James, J. Zeng, D. Smith, A. Domschke, and C.R. Lowe. Holographic glucose sensors. *Biosensors and Bioelectronics*, 20:1602–1610, 2005.

137. A. Domschke, W.F. March, S. Kabilan, and C. Lowe. Initial clinical testing of a holographic non-invasive contact lens glucose sensor. *Diabetes Technology & Therapeutics*, 8(1):89–93, 2006.

138. Y. Hayashi, L. Livshits, A. Caduff, and Y. Feldman. Dielectric spectroscopy study of specific glucose influence on human erythrocyte membranes. *Journal of Physics D: Applied Physics*, 36:369–374, 2003.

139. A. Sieg, R.H. Guy, and M.B. Delgado-Charro. Noninvasive and minimally invasive methods for transdermal glucose monitoring. *Diabetes Technology & Therapeutics*, 7(1):174–197, 2005.

140. A. Caduff, E. Hirt, Y. Feldman, Z. Ali, and L. Heinemann. First human experiments with a novel non-invasive, non-optical continuous glucose monitoring system. *Biosensors and Bioelectronics*, 19:209–217, 2003.

141. G. Velho, PH. Froguel, D.R.Thevenot, and G. Reach. Strategies for calibrating a subcutaneous glucose sensor. *Biomed. Biochim Acta*, 48(11/12):957–964, 1989.

142. U. Fischer, R. Ertle, K. Rebrin, E. Brunstein, H. Hahn von Dorsche, and E.J. Freyse. Assessment of subcutaneous glucose concentration: Validation of the wick in normal technique as a reference for implanted electrochemical sensors in normal and diabetic dogs. *Diabetologia*, 30:940–945, 1987.

143. MiniMed Inc., Sylmar CA. *MiniMed CGMStm continuous glucose monitoring system User's Guide*, March 2000.

144. M.J. Tierney, J.A. Tamada, R.O. Potts, L. Jovanovic, S. Gard, and Cygnus Research Team. Clinical evaluation of the glucowatch(r) biographer: a continual, non-invasive glucose monitor for patients with diabetes. *Biosensors & Bioelectronics*, 16:621–629, 1999.

145. John Pickup. Sensitive glucose sensing in diabetes. *The Lancet*, 355:426–427, 5 February 2000.

146. B. Berner, T.C. Dunn, K.C. Farinas, M.D. Garrison, R.T. Kurnik, M.J. Lesho, R.O. Potts, J.A. Tamada, and M.J. Tierney. Signal processing for measurement of physiological analysis, 2001. United States Patent US 6,233,471 B1.

147. R.T. Kurnik, J.J. Oliver, S.R. Waterhouse, T. Dunn, Y. Jayalakshmi, M. Lesho, M. Lopatin, J. Tamada, C. Wei, and R.O. Potts. Application of the mixtures of experts algorithm for signal processing in a noninvasive glucose monitoring system. *Sensors and Actuators B*, 60:19–26, May 1999.

148. G.M. Steil, K. Rebrin, P.V. Goode, J.J. Mastrototaro, R.E. Purvis, W.P. van Antwerp, and J.J. Shin. Closed loop system for controlling insulin infusion, 2000. World Intellectual Property Organization WO00/74753 A1.

149. William L. Clarke, Daniel Cox, Linda A. Gonder-Frederick, William Carter, and Stephen L. Pohl. Evaluating clinical accuracy of systems for self-monitoring of blood glucose. *Diabetes Care*, 10(5):622–628, 1987.

150. P.D. Home. *Alternative Methods, Systems and Routes of Insulin Delivery*, volume 1, chapter 46, page 446. Blackwell Scientific Publications, 1991.

151. L.L. Young and M.A. Koda-Kimble. *Applied Therapeutics: The clinical use of drugs.*, chapter 48. Applied Therapeutics, Inc., Vancouver WA, 7th edition, 2000.

152. N. Pørksen. The in vivo regulation of pulsatile insulin secretion. *Diabetologia*, 45:3–20, 2002.

153. W.C. Duckworth, C.D. Saudek, and R.R. Henry. Why intraperitoneal delivery of insulin with implantable pumps in NIDDM? *Diabetes*, 41:657–661, June 1992.

154. M.A. Fishel, G.S. Watson, T.J. Montine, Q. Wang, P.S. Green, J.J. Kulstad, D.G.Cook, E.R. Peskind, L.D. Baker, D.Goldgaber, W.Nie, S. Asthana, S.R. Plymate, M.W. Schwartz, and S. Craft. Hyperinsulinemia provokes synchronous increases in central inflammation and β-amyloid in normal adults. *Archives of Neurology*, 62:1539–1544, 2005.

155. J.A. Luchsinger, Ming-Xin Tang, S. Shea, and R Mayeux. Hyperinsulinemia and risk of alzheimer disease. *Neurology*, 63:1187–1192, 2004.

156. I.A. Macfarlane. *The Pharmacokinetics of Insulin*, chapter 38, pages 371–383. Blackwell Scientific Publications, 1991.

157. K.S. Polonsky, M.M. Byrne, and J. Sturis. Alternative insulin delivery systems: How demanding should the patient be? *Diabetologia*, 40:S97–S101, 1997.

158. J.P. Mordes and A.A. Rossini. Management of diabetes in the critically ill patient. In J.M. Rippe, R.S. Irwin, J.S. Alpert, and M.P. Fink, editors, *Intensive Care Medicine 2nd Ed.*, chapter 92, pages 955–962. Little, Brown and Company, 1991.

159. N.M. O'Meara and K.S. Polonsky. Insulin secretion in vivo. In C.R. Kahn and G.C. Weir, editors, *Joslin's Diabetes Mellitus 13th Ed*, chapter 5, pages 81–96. Lea & Febiger, Malvern, Pennsylvania, 1994.

160. D. Elahi. In praise of the hyperglycemic clamp. *Diabetes Care*, 19(3):278–286, 1996.

161. L.C. Ramirez and P. Raskin. *Pancreatic Abnormalities in Non-Insulin-Dependent Diabetes Mellitus*, volume 1, chapter 22, pages 198–204. Blackwell Scientific Publications, 1991.

162. D.S. Bell. Importance of postprandial glucose control. *South Med. J.*, 94(8):804–809, 2001.

163. E.W. Kraegen, D.J. Chisholm, and M.E. McNamara. Timing of insulin delivery with meals. *Horm Metab Res*, 13:365–367, 1981.

164. A. Motluk and L.Geddes. Breakthrough sheds light on cause of diabetes, 15 December 2006. New Scientist Tech [Online] Available at http://www.newscientisttech.com/article.ns?id=dn10812.

165. B.C. Borlase, T.J. Babineau, R.A. Forse, S.J. Bell, and G.L. Blackburn. Enteral nutritional support. In J.M. Rippe, R.S. Irwin, J.S. Alpert, and M.P. Fink, editors, *Intensive Care Medicine 2nd Ed.*, chapter 174, pages 1669–1674. Little, Brown and Company, 1991.

166. D. Dazzi, F. Taddei, A. Gavarini, E. Uggeri, R. Negro, and A. Pezzarossa. The control of blood glucose in the critical diabetic patient: A neuro-fuzzy method. *Journal of Diabetes and Complications*, 15:80–87, 2001.

167. F. Chee, T. Fernando, and P.V. van Heerden. Closed-loop glucose control in critically-ill patients using continuous glucose monitoring system (CGMS) in real-time. *IEEE Trans. Inf. Tech. Biomed.*, 7(1):43–53, March 2003.

168. E. R. Carson, C. Cobelli, and L. Finkelstein. *The Mathematical Modelling of Metabolic and Endocrine Systems - Model Formulation, Identification, and Validation*, volume 2 of *Biomedical Engineering and Health Systems*. John-Wiley and Sons, 1983.

169. Y. Cherruault. *Mathematical Modelling in Biomedicine*. D. Reidel Publishing Company, Dordrecht, Holland, 1986.

170. J.C. Henquin. Cell biology of insulin secretion. In C. R. Kahn and G.C. Weir, editors, *Joslin's Diabetes Mellitus 13th Ed*, pages 56–80. Lea & Febiger, Malvern, Pennsylvania, 1994.

171. R.A. Rizza, L.J. Mandarino, and J.E. Gerich. Dose-response characteristics for effects of insulin on production and utilization of glucose in man. *Am. J. Physiol. (Endocrinol. Metab)*, 240:E630–E639, 1981.

172. A.M. Albisser, B.S. Leibel, T.G. Ewart, Z. Davidovac, C.K. Botz, and W. Zingg. An artificial endocrine pancreas. *Diabetes*, 23(5):389–396, May 1974.

173. E.W. Kraegen, R. Whiteside, D. Bell, Y.O. Chia, and L. Lazarus. Development of a closed-loop artificial pancreas. *Horm Metab Res Suppl*, 8:39–42, 1979.

174. J.D. Kruse-Jarres, G. Braun, R. Naegeie, M. Bresch, and U. Lehmann. Blood glucose monitoring and computer regulation by means of an artificial endocrine pancreas. *Horm Metab Res Suppl*, 8:43–45, 1979.

175. N.H. White, D. Skor, and J.V. Santiago. Practical closed-loop insulin delivery: A system for the maintenance of overnight euglycemia and the calculation of basal insulin requirements in insulin-dependent diabetes. *Ann Intern Med*, 97:210–213, 1982.

176. S.M. Furler, E.W. Kraegen, R.H. Smallwood, and D.J. Chisolm. Blood glucose control by intermittent loop closure in the basal mode: Computer simulation studies with a diabetic model. *Diabetes Care*, 8(6):553–561, 1985.

177. R.L. Ollerton. Application of optimal control theory to diabetes mellitus. *Int. J. Control*, 50(6):2503–2522, 1989.

178. P. Palazzo and V. Viti. A new glucose-clamp algorithm theoretical consideration and computer simulations. *IEEE Trans. Biomed. Eng.*, 37(5):535–543, 1990.

179. F. Chee, T.L. Fernando, A.V. Savkin, and P.V. van Heerden. Expert PID control system for blood glucose control in critically-ill patients. *IEEE Trans. Inf. Tech. Biomed.*, 7(4):419–425, December 2003.

180. G.M. Steil, A.E. Panteleon, and K. Rebrin. Closed-loop insulin delivery - the path to physiological glucose control. *Advanced Drug Delivery Reviews*, 56:125–144, 2004.

181. E. Renard, G. Costalat, H. Chevassus, and J. Bringer. Closed loop insulin delivery using implanted insulin pumps and sensors in type 1 diabetic patients. *Diabetes Research and Clinical Practice*, 74:S173–S177, 2006.

182. B. Gopakumaran, H.M. Duman, D.P. Overholser, I.F. Federiuk, M.J. Quinn, M.D. Wood, and W.K. Ward. A novel insulin delivery algorithm in rats with type 1 diabetes: The fading memory proportional-derivative method. *Artifical Organs*, 29(8):599–607, 2005.

183. G. Marchetti, M. Barolo, L. Jovanovic, and H. Zisser. An improved PID switching control strategy for Type 1 diabetes. In *Proceedings of the 28th IEEE EMBS Annual International Conference*, pages 5041–5044, New York City, USA, 30 August-3 September 2006. IEEE.

184. V.W. Bolie. Coefficients of normal blood glucose regulation. *Journal of Applied Physiology*, 16(5):783–788, 1961.

185. E. Ackerman, J.W. Rosevear, and W.F. McGuckin. A mathematical model of the glucose tolerance test. *Physics in Medicine and Biology*, 9(2):203–213, 1964.

186. E. Ackerman, Laël.C. Gatewood, J.W. Rosevear, and G.D. Molnar. Model studies of blood-glucose regulation. *Bulletin of Mathematical Biophysics*, 27:21–37, 1965. Special Issue.

187. L.C. Gatewood, E. Ackerman, J.W. Rosevear, and G.D. Molnar. Test of a mathematical model of the blood-glucose regulatory system. *Computers and Biomedical Research*, 2:1–14, 1968.

188. F. Ceresa, F. Ghermi, P. F. Martini, P. Martino, G. Segre, and A. Vitelli. Control of blood glucose in normal and in diabetic subjects - studies by compartmental analysis and digital computer technics. *Diabetes*, 17(9):571–578, 1968.

189. T. Chorbajian. An in numero study of glucose-insulin interaction. *Bulletin of Mathematical Biophysics*, 33:451–462, 1971.

190. E. Cerasi, G. Fick, and M. Rudemo. A mathematical model for the glucose induced insulin release in man. *European Journal of Clinical Investigation*, 4:269–278, 1974.

191. M. Nomura, M. Shichiri, R. Kawamori, Y. Yamasaki, N. Iwama, and H. Abe. A mathematical insulin-secretion model and its validation in isolated rat pancreatic islets perifusion. *Computers and Biomedical Research*, 17:570–579, 1984.

192. J. S. Bajaj, G. Subba Rao, J. Subba Rao, and R. Khardori. A mathematical model for insulin kinetics and its application to protein-deficient (malnutrition-related) diabetes mellitus (PDDM). *Journal of Theoretical Biology*, 126:491–503, 1987.

193. G. Subba Rao, J. S. Bajaj, and J. Subba Rao. A mathematical model for insulin kinetics II. extension of the model to include response to oral glucose administration and application to insulin-dependent diabetes mellitus (IDDM). *Journal of Theoretical Biology*, 142:473–483, 1990.

194. L. Jansson, L. Lindskog, N. E. Norden, S. Carlstrom, and B. Schersten. Diagnostic value of the oral glucose tolerance test evaluated with a mathematical model. *Computers and Biomedical Research*, 13:512–521, 1980.

195. E. Salzsieder, G. Albrecht, U. Fischer, and E.-J Freyse. Kinetic modeling of glucoregulatory system to improve insulin therapy. *IEEE Trans. Biomed. Eng.*, 32(10):846–855, October 1985.

196. R.N. Bergman and J. Urquhart. The pilot gland approach to the study of insulin secretory dynamics. *Recent Prog Horm Res*, 27:583–605, 1971.

197. R. N. Bergman, L. S. Phillips, and C. Cobelli. Physiological evaluation of factors controlling glucose tolerance in man. *Journal of Clinical Investigation*, 68:1456–1467, December 1981.

198. G. Toffolo, R. N. Bergman, D. T. Finegood, C. R. Bowden, and C. Cobelli. Quantitative estimation of beta cell sensitivity to glucose in the intact organism - a minimal model of insulin kinetics in the dog. *Diabetes*, 29:979–990, December 1980.

199. B. Candas and J. Radziuk. An adaptive plasma glucose controller based on a nonlinear insulin/glucose model. *IEEE Trans. Biomed. Eng.*, 41(2):116–124, February 1994.

200. F.J. Doyle III, C. Dorski, J. Harting, and N.A. Peppas. Control and modelling of drug delivery devices for the treatment of diabetes. In *Proceedings of the American Conference*, volume 1, pages 776–780, Seattle, Washington, USA, 21-23 June 1995. IEEE.

201. A. De Gaetano and O. Arino. Mathematical modeling of the intravenous glucose tolerance test. *Journal of Mathematical Biology*, 40:136–168, 2000.

202. C. Cobelli, G. Federspil, G. Pacini, A. Salvan, and C. Scandellari. An integrated mathematical model of the dynamics of blood glucose and its hormonal control. *Mathematical Biosciences*, 58:27–60, 1982.

203. C. Cobelli and A. Mari. Control of diabetes with artificial systems for insulin delivery - algorithm independent limitations revealed by a modeling study. *IEEE Trans. Biomed. Eng.*, 32(10):840–845, October 1985.

204. C. Cobelli, G. Pacini, G. Toffolo, and L. Sacca. Estimation of insulin sensitivity and glucose clearance from minimal model: new insights from labeled IVGTT. *American Journal of Physiology*, 250:E591–E598, 1986.

205. D. K. Parrish and D. B. Ridgely. Control of artificial human pancreas using the SDRE method. In *Proceedings of the American Control Conference*, pages 1059–1060, Albquerque, New Mexico, June 1997. Academic Press.

206. R. S. Parker, F. J. Doyle III, and N. A Peppas. A model-based algorithm for blood glucose control in type I diabetic patients. *IEEE Trans. Biomed. Eng.*, 46(2):148–157, February 1999.

207. R.S. Parker, F.J. Doyle III, J.H. Ward, and N.A. Peppas. Robust H^∞ glucose control in diabetes using a physiological model. *AIChE Journal*, 46(12):2537–2549, December 2000.

208. E. RuizVelázquez, R. Femat, and D.U. CamposDelgado. Blood glucose control for type I diabetes mellitus: A robust H^∞ tracking problem. *Control Engineering Practice*, 12:1179–1195, 2004.

209. G.M. Steil, B. Clark, S. Kanderian, and K. Rebrin. Modelling insulin action for development of close-loop artificial pancreas. *Diabetes Technology & Therapeutics*, 7(1):94–108, 2005.

210. R. Hovorka, F. Shojaee-Moradie, P.V. Carroll, L.J. Chassin, I.J. Gowrie, N.C. Jackson, R.S. Tudor, A.M. Umpleby, and R.H. Jones. Partitioning glucose distribution/transport, disposal and endogenous production during IVGTT. *Am J Physiol Endocrinol Metab*, 282:992–1007, 2002.

211. R. Hovorka, V. Canonico, L.J. Chassin, U. Haueter, M. Massi-Benedetti, M.O. Federici, T.R. Pieber, H.C. Schaller, L. Schaupp, T. Vering, and M.E. Wilinska. Nonlinear model predictive control of glucose concentration in subjects with type 1 diabetes. *Physiological Measurement*, 25:905–920, 2004.

212. M. Giugliano, M. Bove, and M. Grattarola. Insulin release at the molecular level: Metabolic-electrophysiological modeling of the pancreatic beta-cells. *IEEE Eng. Med. Biol. Mag.*, 47(5):611–623, May 2000.

213. K. Ogata. *Modern Control Engineering*. Prentice-Hall International, Inc, Upper Saddle River, New Jersey, 3rd edition, 1997.

214. U. Fischer, W. Schenk, E. Salzsieder, G. Albrecht, P. Abel, and E.-J Freyse. Does physiological blood glucose control require an adaptive control strategy? *IEEE Trans. Biomed. Eng.*, 34(8):575–582, Aug 1987.

215. C. Dalla Man, G. Toffolo, and C. Cobelli. Gluocse kinetics during meal:one vs two compartment minimal model. In *Proceedings of the Second Joint EMBS/BMES Conference*, pages 2222–2223, Houston, TX, USA, 23 -26 October 2002. IEEE.

216. C. Dalla Man, R.A. Rizza, and C. Cobelli. Mixed meal simulation model of glucose-insulin system. In *Proceedings of the 28th IEEE EMBS Annual International Conference*, pages 307–310, New York City, USA, 30 August-3 September 2006. IEEE.

217. J.T. Sorensen, C.K. Colton, R.S. Hillman, and J.S. Soeldner. Use of a physiologic pharmacokinetic model of glucose homeostasis for assessment of performance requirements for improved insulin therapies. *Diabetes Care*, 5(3):148–157, 1982.

218. M. Ader, G. Pacini, Y.J. Yang, and R. Bergman. Importance of glucose per se to intravenous glucose tolerance. *Diabetes*, 34:1093–1103, 1985.

219. F. J. Vega-Catalan. A program to estimate insulin sensitivity and pancreatic responsivity from an IVGTT using the minimal modeling technique. *Computers and Biomedical Research*, 23:1–9, 1990.

220. T. Van Herpe, B. Pluymers, M. Espinoza, G. Van den Berghe, and B. De Moor. A minimal model for glycemia control in critically ill patients. pages 5432–5435. IEEE, 30 August-3 September 2006.

221. R. Hovorka, L.J. Chassin, M.E. Wilinska, V. Canonico, J.A. Akwi, M.O. Federici, M. Massi-Benedetti, I. Hutzli, C. Zaugg, H. Kaufmann, M. Both, T. Vering, H.C. Schaller, L. Schaupp, M. Bodenlenz, and T.R. Pieber. Closing the loop: The adicol experience. *Diabetes Technology & Therapeutics*, 6(3):307–318, 2004.

222. J.T. Sorensen. *A Physiological Model of Glucose Metabolism in Man and it Use to Design and Access Improved Insulin Therapies for Diabetes*. PhD thesis, Dept.of Chemical Engineering, MIT, 1985.

223. S.M. Lynch and B.W. Bequette. Model predictive control of blood glucose in type i diabetes using subcutaneous glucose measurements. In *Proceedings of the American Control Conference*, Anchorage, AK, 8 -10 May 2002.

224. T.L Yates and L.R. Fletcher. Prediction of a glucose appearance function from food using deconvolution. *IMA Journal of Mathematics Applied in Medicine and Biology*, 17:169–184, 2000.

225. Z. Trajanoski and P. Wach. Neural predictive controller for insulin delivery using the subcutaneous route. *IEEE Trans. Biomed. Eng.*, 45(9):1122–1134, 1998.

226. N.L. Ricker. Model predictive control with state estimation. *Ind. Eng. Chem. Res*, 29:374–382, 1990.

227. M. E. Fisher and K. L. Teo. Optimal insulin infusion resulting from a mathematical model of blood glucose dynamics. *IEEE Trans. Biomed. Eng.*, 36(4):479–486, April 1989.

228. G.W. Swan. An optimal control model of diabetes mellitus. *Bulletin of Mathematical Biology*, 44(6):793–808, 1982.

229. M. Gopal. *Modern Control System Theory*. Wiley Eastern Ltd, New Delhi, 1984.

230. Anthony Zaknich. *Principles of Adaptive Filters and Self-learning Systems*. ISBN 1-85233-984-5. Springer-Verlag, 2005. Series on Advanced Textbooks in Control and Signal Processing.

231. B. Pagurek, J.S. Riordon, and S. Mahmoud. Adaptive control of the human glucose-regulatory system. *Med. & Biol. Eng*, 10:753–761, 1972.

232. M. Kikuchi, E. Machiyama, N. Kabei, and A. Yamada. Adaptive control system of blood glucose regulation. In Lindberg/Kaihara, editor, *MEDINFO 80*, pages 96–100. IFIP, North-Holland Publishing Company, 1980.

233. E. Sarti and P. Cruciani. Self-tuning control algorithm for wearable artificial pancreas. In *Vol. 14. Proceedings of the Annual International Conference of the IEEE*, volume 6, pages 2267–2269. Engineering in Medicine and Biology Society, 29 October-1 November 1992.

234. M. Morari and N.L. Ricker. *Model Predictive Control Toolbox User's Guide*. The Mathworks, Inc, Natick, MA, 1998.

235. C.E. García, D.M. Prett, and M. Morari. Model predictive control: Theory and practice - a survey. *Automatica*, 25(3):335–348, 1989.

236. R. Bellazzi, G. Nucci, and C. Cobelli. The subcutaneous route to insulin-dependent diabetes therapy. *IEEE Eng. Med. Biol. Mag.*, 20(1):54–64, January-February 2001.

237. R.T. Stefani, B. Shahian, C.J. Savant, and G.H. Hostetter. *Design of Feedback Control Systems*. Oxford University Press, New York, 4th edition, 2002.

238. K. Glover and J.C. Doyle. State-space formulae for all stabilizing controllers that satisfy and H_∞ -norm bound and relations to risk sensitivity. *System & Control Letters*, 11:167–172, 1988.

239. W.S. Levine and R.T. Reichert. An introduction to H_∞ control system design. In *Proceedings of the 29th Conference on Decision and Control*, Honolulu, Hawaii, December 1990.

240. J.C. Doyle, K. Glover, P.P. Khargonekar, and B.A. Francis. State-space solutions to standard H_2 and H_∞ control problems. *IEEE Transactions on Automatic Control*, 34(8):831–847, 1989.

241. H. Toivonen. Robust control method, AS-74.330 Postgraduate Course in Control Engineering, 2002. Process Control Laboratory, Abo Akademi [Online] Available at http://www.abo.fi/~htoivone/courses/AS74330.html.

242. K. H. Kienitz and T. Yoneyama. A robust controller for insulin pump based on H^∞ theory. *IEEE Trans. Biomed. Eng.*, 40(11):1133–1137, December 1993.

243. F. Chee, A.V. Savkin, T.L. Fernando, and S. Nahavandi. Optimal H^∞ insulin injection control for blood glucose regulation in diabetic patients. *IEEE Trans. Biomed. Eng.*, 52(10):1625–1631, October 2005.

244. A.V. Savkin and R.J. Evans. A new approach to robust control of hybrid systems over infinite time. *IEEE Transactions on Automatic Control*, 43(9):1292–1296, 1998.

245. A.V. Savkin and I.R. Petersen. Robust filtering and model validation for uncertain continuous-time systems with discrete and continuous measurements. *International Journal of Control*, 69(1):163–174, 1998.

246. A.V. Savkin and R.J. Evans. *Hybrid Dynamical Systems. Controller and Sensor Switching Problems*. Birkhäuser, Boston, Cambridge MA, 1st edition, 2002.

247. D.P. Bertsekas and I.B. Rhodes. Recursive state estimation for a set-membership description of uncertainty. *IEEE transactions on automatic control*, AC-16(2):117–128, April 1971.

248. D.A. Gough, S. Aisenberg, C.K. Colton, J.Giner, and J.S. Soeldner. The status of electrochemical sensors for in vivo glucose monitoring. *Horm Metab Res Suppl*, 7:10–22, 1977.

249. R.D. Beach, F. V. Kuster, and F. Moussy. Subminiature implantable potentiostat and modified commercial telemetry device for remote glucose monitoring. *IEEE Transaction in Instrumentation and Measurement*, 48(6):1239–1245, 1999.

250. K.C. McIvor, J.L. Cabernoch, N.M. Van Antwerp, E.C. Halili, and J.J. Mastrototaro. Glucose sensor package system, 2002. US Patent 6,360,888 B1.

251. S.J. Updike, M.C. Shults, B.J. Gilligan, and R.K. Rhodes. A subcutaneous glucose sensor with improved longevity, dynamic range, and stability of calibration. *Diabetes Care*, 23(2):208–214, 2000.

252. T. Yipintsoi, L.C. Gatewood, E. Ackerman, P.L. Spivak, G.D. Molnar, J.W. Rosevear, and F.J. Service. Mathematical analysis of blood glucose and plasma insulin responses to insulin infusion in healthy and diabetic subjects. *Computers in biology & medicine*, 3:71–78, 1973.

Index

Printing: Mercedes-Druck, Berlin
Binding: Stein+Lehmann, Berlin

Lecture Notes in Control and Information Sciences

Edited by M. Thoma, M. Morari

Further volumes of this series can be found on our homepage:
springer.com